Visual Ergonomics for Communication Design

Visual Ergonomics for Communication Design reveals the application of ergonomics principles in visual communication design. It enables the visual designer to look at different aspects of visual design from the point of view of the user to ensure that their designs are more user friendly.

The book allows the reader to apply the principles of ergonomics in different facets of communication design such as pictograms, icons and logo design, product labelling, information systems in spaces, and visual ergonomics in simple map design. An introductory chapter allows the reader to learn the basics and principles of visual ergonomics and gives them an insight into the probable application areas. Further chapters delve deeper into the topic with each chapter ending with "Key Points", "Practice Session", and exercises designed to help the reader to revise what they have learned. The reader will benefit from coverage of visual ergonomics in icons, pictograms, and symbols, product labelling, wayfinding in spaces, and map design which are all backed up with examples and illustrations. Written with the layperson in mind who has no background in the subject, this interesting and easy-to-read reference addresses how to use the different principles of ergonomics in visual communication in a storytelling format.

With a narrative structure to the chapters and illustrations throughout, this book is an ideal read for students and professionals looking to strengthen their knowledge of visual communication design by augmenting it with ergonomic principles. It will appeal to students and professionals studying and working in the fields of computer science, human–computer interaction, design engineering, mechanical engineering, information technology, communications engineering, and human factors engineering.

Visual Ergonomics for Communication Design

A Layperson's Approach

Prabir Mukhopadhyay
MSc, PhD

CRC Press
Taylor & Francis Group
Boca Raton London New York

CRC Press is an imprint of the
Taylor & Francis Group, an **informa** business

First edition published 2023
by CRC Press
6000 Broken Sound Parkway NW, Suite 300, Boca Raton, FL 33487-2742

and by CRC Press
4 Park Square, Milton Park, Abingdon, Oxon, OX14 4RN
CRC Press is an imprint of Taylor & Francis Group, LLC

ISBN: 9781032436876 (hbk)
ISBN: 9781032439419 (pbk)
ISBN: 9781003369516 (ebk)

DOI: 10.1201/9781003369516

Typeset in Times
by Deanta Global Publishing Services, Chennai, India

To the Lotus Feet of Sri Rama Krishna Param Hansha Dev …

Late Dhirendra Nath Mukhopadhyay (Father), miss you a lot!!

&

Mrs. Meena Mukhopadhyay (Mother), Dada, Mamoni,
Duggi, and Leto …

Contents

Preface

At the outset I would clarify that this book is not on the design of computer workstations, nor does it deal with the different diseases related to prolonged computer usage. This is a book which discusses visual ergonomic aspects of communication design like icons, symbols, pictograms, product labelling and packaging, wayfinding in spaces, just to name a few.

Communication design is a multidisciplinary field today. We have students from science and technology background and we also have a large number of students from arts, humanities, literature, etc. as well. This diverse background of students makes it quite challenging when it comes to introducing the subject "visual ergonomics". This is because one has to know the subject in depth which is deeply rooted in science and technology in order to be able to apply it in different facets of communication design.

This book is an attempt in that direction that it touches upon the different aspects of visual ergonomics in communication design (mainly visual communication), and in an easy to understand, layperson's language. The book is written in a storytelling format without using any technical jargon so that literally anyone irrespective of his or her background would be able to read this book and apply the visual ergonomic principles in communication design. Each chapter starts with an overview of what the chapter is all about and ends with some practice sessions so that readers can learn to apply them. The chapters touch upon different aspects of communication design. Chapter 6 comprises a set of assignments with design directions. The readers should attempt this chapter after reading the entire book. I hope the storytelling format in the book would ensure that readers are not bored as they read the different chapters, and would be able to relate the concepts in their daily life. The images used in this book are for representation purposes (except for the line diagrams) and have been used to build the story line and make the content interesting to the readers.

Again, I wish to express my gratitude to those people who have helped in preparing this book. I am particularly grateful to James Hobbs, Kirsty Hardwick, and all the staff of CRC Press, Taylor and Francis Group, for their continued editorial and production support all through the publication process.

Acknowledgements

Acknowledgement is due to many people for helping me in completing this book.

The following students have worked hard towards data collection, analysis, and concept generation, and I owe them a lot.

Chapter 4: Figures 4.1 to 4.6: Gaurav Patel and Shalabh Singh, MDes student, Design Discipline, Indian Institute of Information Technology Design and Manufacturing, Jabalpur, India.

Figures 4.7 to 4.12: Bipratim Saha, MDes student, Design Discipline, Indian Institute of Information Technology Design and Manufacturing, Jabalpur, India.

Chapter 5: Figure 5.2 to 5.8: Arushi Choudhary, BDes student, Design Discipline, Indian Institute of Information Technology Design and Manufacturing, Jabalpur, India.

Figure 5.11 to 5.17: Kanti Upadhay, Arjun Hembrum, and Bhanu Kiran, MDes student, Design Discipline, Indian Institute of Information Technology Design and Manufacturing, Jabalpur, India.

My PhD scholar Gaurav Patel has helped in fine tuning many of the figures and has contributed to Figures 6.1 and 6.2.

A very special thanks to Mr Vipul Vinzuda, Faculty of Transportation and Automobile Design, Post Graduate Campus, National Institute of Design, Gandhinagar, India, for inspiring me to write this book and helping me in fine tuning many of the figures at a very short notice.

Author

Dr Prabir Mukhopadhyay
Associate Professor and Head, Design Discipline,
Indian Institute of Information Technology
Design and Manufacturing, Jabalpur, India.

Prabir Mukhopadhyay holds a BSc Honours Degree in Physiology and an MSc Degree in Physiology with specialisation in Ergonomics and Work Physiology both from Calcutta University, India. He holds a PhD in Industrial Ergonomics from the University of Limerick, Ireland. Prabir started working with noted ergonomist Prof R.N. Sen, at Calcutta University both for his master's thesis and later on a project sponsored by the Ministry of Environment and Forests, Government of India. It was during this time Prabir developed a keen interest in the subject and wanted to build his career in ergonomics. He joined the National Institute of Design, Ahmedabad, India, as an ergonomist for one of the projects for the Indian Railways. There he was mentored by Dr S. Ghosal, the project lead. He then joined the same institute as a faculty in ergonomics. During his tenure at Ahmedabad, he worked on many consultancy projects related to ergonomics. Some of his clients there included the Indian Railways, Self Employed Women's Association, and the United Nations Industrial Development Organization.

After working there for two years, Prabir left for the University of Limerick, Ireland, on a European Union funded project under the supervision of Prof T.J. Gallwey. He completed his PhD in Industrial Ergonomics from the same university and decided to return to India to apply his acquired knowledge. He joined the National Institute of Design, Post Graduate Campus at Gandhinagar, India, as a faculty in ergonomics. There he headed the Software and User interface Design discipline. He also completed a research project funded by Ford Foundation-National Institute of Design on ergonomics design intervention in the craft sectors at Jaipur in Rajasthan, India. Simultaneously, he started teaching ergonomics across different design disciplines at other campuses of the institute like Ahmedabad and Bangalore as well.

After working there for around five years, Prabir joined his present institute as an assistant professor in Design. He was later promoted to an associate

professor and later became the Discipline Head. He teaches practices and research in different areas of ergonomics and its application in design. He has authored three books to date: *Ergonomics for the Layman: Application in Design*, published by CRC Press in 2019, *Ergonomics Principles in Design: An Illustrated Fundamental Approach*, published by CRC Press in 2022, and *Ergonomics in Fashion Design: A Layperson's Approach*, published by Springer, Singapore in 2022. He is a bachelor, and his hobbies include watching action movies, listening to Indian and Western music, travelling, and cooking.
Contact: prabirdr@gmail.com

The World of Wonders, Origin, and How to Move Forward

1

Overview

This chapter is an introduction to the field of visual ergonomics, its components and application areas, and how it can be used as a power medium in different spheres and domains of our life. The storyline in the chapter takes the reader through different nuances of the subject and where and how it could be applied to make users' lives much more comfortable and satisfying. This chapter gives an insight into the different principles of visual ergonomics which one needs to keep at the back of their mind. The chapter discusses the importance of the users and the context of use when it comes to applying the different principles of visual ergonomics. Some ordinary and extraordinary circumstances under which visual ergonomics aids in effective communication are also discussed in this chapter.

In the entire book, the word information has been used as a very general term to indicate what is communicated to the user. It may please not be taken as a technical definition for information.

1.1 INTRODUCTION

There was a wise man, and people used to go to him for advice. His name was Mr Honest, who used to travel a lot, enjoy nature. He had the habit of interacting with new products, places, and people. Every time he was out to venture into the known and the unknown, he learnt something new and tried to share

DOI: 10.1201/9781003369516-1

that with others. He was an avid reader, a writer, and an artist. To him life was all about people, products, spaces around him and he thoroughly enjoyed them. One day, he got up in the morning and was unable to see anything. He could feel the warmth of the sun but could not see the world around him. He was taken to the doctor and after many months of treatment finally got his eyesight back. He was now extremely happy and at the same time realised the most precious gift of God, the "eyes". This is where we stand today. We do not realise how important our eyes are until we have problems with them. The eyes are the major avenues through which communication between the world around us and ourselves is established. We need to know about this "window" through which we communicate with the beautiful world: products, spaces, people and their love and affection are augmented by the eyes. Thus, it's important that we know about this window and how it works and its major features.

1.2 THE FABULOUS WORLD AROUND US

We see and we believe what we see. We look at products, spaces, devices around us, appreciate them, and use them. Visual apparatus or the eyes help us in this pursuit. Ergonomics as a subject helps us in making our lives comfortable and more meaningful. This is achieved by designing products, spaces, communication, and services in tandem with human needs, wants, and capabilities. Visual ergonomics is the ergonomics of eyes, mainly dealing with how, what, where we see things and how this communication between the eyes and the world around us can be improved for augmenting human communication between products, spaces, and services. We go for a movie and enjoy it because we have our eyes through which we see things. We go to a shop to buy another mobile phone; we appreciate its colour, form, and the different features. We travel on the road and look at different signals, signage, etc. which help us in making our journey more fruitful and at times enjoyable and provide us with a lifetime of memories.

In this world, visual ergonomics equips us to design everything around us which has direct communication with the eyes, in a manner that things become easy, enjoyable, and productive for us. If there were no traffic lights, we would have met with accidents. If there were no road signage, we would have to ask people while moving down the roads. If our mobile phones did not have those icons, we would have no clue how to operate the device. How about your microwave oven with only switches and no other communications as to how to operate?

The vulture and the Himalayan goose are two birds that fly high in the sky but with different objectives. The vulture while flying high in the sky focuses only on dead bodies on the earth below. The Himalayan goose on the other hand looks at the world below, enjoys the beauty of nature, the paddy fields, rivers, mountains, people, and other places. For both, the eyes are important but they use them for two different purposes. The vision and intention of the goose is to look into the earth below in totality and the individual elements like mountains, rivers, paddy fields which make up this total picture. This is where the eyes are important in that they have to be used in a manner so that one could reap the maximum benefit out of them.

1.3 OUR SENSES AND THEIR ROLE

There are a couple of sense organs in our body which act as receivers of information from the world around us. These sense organs are the eyes, ears, nose, tongue, and skin. Of these, the eyes and the ears are very good friends and it's prudent to keep them together. The nose and the tongue are again good friends and they work together. If you have a bad cold and suffer from nasal blockage, you will notice that you do not enjoy your food that day. Most of these senses are located at the topmost point in the body (head region) just like a mobile tower so that receiving signals from the outside world becomes a little easier. Similarly, they are located closer to the head or the command-and-control centre of the body that is the brain, why? Who does not want to remain closer to one's boss! Exchange of information in that case becomes very easy. So, nature has very carefully designed and crafted our senses or receivers in the body and a little insight into them becomes handy if we have to design different visual elements around us through which we tend to communicate a plethora of information to our users. This is what designers need to keep in mind while designing any communication system like signage on roads, icons for your favourite multimedia device, symbols for those driving down the road. In all these cases, it's your eyes and the way they see determines whether your users are able to understand what you are trying to convey to them.

When you are in an argument with your friend, you often have to hear your friend saying, "you don't have any sense man". What does this mean? This means that you have some problem with your sensory channels and are not able to receive information from the outside world. Thus, by your senses I mean all the five senses that are the windows through which you are kept aware of what is happening in the world around you and then your brain decides what you should do next. Thus, the senses or the windows to our body and mind essentially decide which information to permit in and then send them to the brain for further synthesis and analysis and decide upon the next course of action. Just imagine that you enter a supermarket in search of a good after-shave lotion. You enter the store and start "looking" around for the section which sells the same. Then you come closer to the rack and start "looking" at different brands and finally pick one. You then read the ingredients and if there is a sample bottle open, you try to take some in your hand and smell. After this, you take a decision whether to buy or not to buy. Your eyes and nose are now working together to help you take a decision and buy a product. In fact, your senses work together as and when the situation demands. Here two of your senses worked together and this is how all our senses work. Our focus of attention would be mainly on the eyes here, and as we talk about the eyes, we would touch upon other senses as and when required.

1.4 EYES THE MOST PRECIOUS GIFT OF NATURE

You are familiar with a camera I believe. I am not talking about the digital camera only or the camera in your phones available nowadays. I am talking about those film cameras that were once very popular among the older generation. So, what does a camera do? It captures different elements around us, ranging from people, places, products, nature, and the list goes on. In other words, the camera helps us to "record" things around us so we might refer to them later. Our eyes have also some similarities with the film camera. Let us now have a look at the eyes and the camera in a comparative manner.

A camera is portable and you can carry it anywhere you like. The eyes are more than that. They are attached to your head region and move along with you and do not part even for a second! The camera can fall from your hand and break! Now let us have a look at the internal mechanism of the eyes. Light enters the camera through a small opening in the front called the "aperture" which you can adjust as per your requirements. If the ambient illumination level is very low or it's dark, then you need to change the aperture of the camera to permit more light inside so that you get a clear photograph. The eyes also have a similar mechanism where the light enters through a small hole in the front called the pupil. The size of the pupil grows big and small, allowing more and less light to enter. Compared to the camera, this aperture controlling mechanism in the human eye is automatic and not under your control. In a camera, it's under your control and in an automatic camera, it's done automatically. After the light enters the camera, it has to pass through a lens or at times through multiple lenses. The lenses move back and front and bring any image to focus. So, the focal length in the camera (focal length equals to half of the radius of curvature) is adjusted by backward and frontward movement of the lenses and thus the image is brought into focus. The eyes are very small and are provided with one convex lens but with no room for movement in the forward and backward directions. So how does it focus? How is the radius of curvature changed? Here nature proves itself superior to humans. The convex lenses in the human eyes are jelly like and held in position in the front by a group of muscles called the ciliary muscles. This jelly like lens is pulled by the ciliary muscles and becomes a thin convex lens. With the thinning of the lens its radius of curvature increases and focal length changes. In the same way, when the ciliary muscles relax, the lens thickens and its radius of curvature decreases, thus changing the focal length. This way the eye lens focuses on any object. In the camera, the light hits the photographic plate and chemical changes happen there due to which images are formed. In the eye, the photographic plate is the retina on which image formation takes place and it's in colour. This is because there are some colour specific cells called the "cone cells". Thus, the retina is the equivalent of the photographic plate in a camera. Isn't this structure of the eye amazing?

1.5 ERGONOMICS AND VISUAL ERGONOMICS: WHERE DO THEY GEL

Most of the students make the mistake of relating visual ergonomics with application of ergonomic principles in designing a computer workstation. In reality it's much beyond that and a computer workstation just occupies a small fragment of it which we are not going to discuss at all. Ergonomics in a layperson's language is the relationship between human, artificial products and the environment, and the context of use. Visual ergonomics is the application of ergonomic principles in facilitating communication between humans and a product.

Once upon a time, Ram, a young man in his thirties, travelled to a beautiful country. When he arrived in that country, he could not understand the language of the people there. This was because he was born and brought up in a completely different country. As he was walking down the street, he felt hungry and so he started looking for some restaurants where he could go and eat. He was unable to understand what was written and was a little worried because he was a vegetarian. So, after looking at the picture of a palm tree, he entered one restaurant with the hope that it served vegetarian food. His logic was that if non-vegetarian food was served they would have definitely displayed some pictures of meat or meat slices. As he entered, he was greeted in a foreign language and he took a seat. He took the menu card and had no clue as to what the food was.

The waiter also didn't know his language. So, he started to look at the small visuals next to the food and finally zeroed in on a dish which he thought might be vegetarian because it depicted green salads. He ordered and the food was served. When he looked at the food, he found some brown disc-like product and plenty of green salads. He was happy and thought those brown disc-like products were possibly the root of a plant. He started eating but could smell something to which he was not accustomed to before. He finished his food and paid the bill and as he was about to leave the restaurant, he met a person from his country. They talked for a while and Ram asked "can you let me know what vegetable this is?" Saying this, Ram took the menu card and pointed his finger to the dish that he had just now. The gentlemen told him that's known as "pork chop" and it's served with salads. Ram was stunned. He had misunderstood the visuals and had eaten non-vegetarian food for the first time in his life!

This is exactly where visual ergonomics can help. It ensures the visual communication between users and coded visuals (icons) such that users understand them easily and there is no need for any text, and in such cases, it becomes language independent and easy to decode. The symbol of a red cross indicates a hospital to all the spectrum of population and users irrespective of any location on the globe can understand it very easily. This is what visual ergonomics strives at in making communication between people, spaces, and products easy and ensuring there is minimal error in understanding.

1.6 IT'S BETTER TO KNOW ABOUT WHERE WE CAME FROM

I know that it's difficult to admit that our ancestors (thousands of years before) were monkeys. But that's the reality and we cannot help. In the course of evolution, the human race evolved and gradually became bipedal (we walk on two legs) and had a very well-developed brain. Because of these features we are so proud and we claim to be the best species among the animal kingdom. But humans as a species did not lead their lives as we do today. There were no cell phones, fast food, jobs in the information technology sector, and in other words, it was a very slow life where the needs and wants of humans were different. Our ancestors were believed to hunt and gather their food, eat, and sleep. Quite a cosy life, isn't it? Nature decides the evolution of different species including humans, and thus the body was designed in a way that humans could do their job properly and efficiently, that is, hunting and gathering food. So far, it had been a very peaceful life for humans without much of a problem. The eyes for example were developed to equip humans in a manner that they were able to see distant objects like wild animals and get alerted. In those times, at sunset, humans used to go to sleep and there was no television for them to watch at night! Thus, eyes were designed for seeing during the day and to sustain essential activities at night.

Unfortunately, humans gradually changed their purpose of life. The wheel was invented and an industrial revolution took place. With the industrial revolution, the purpose of humans which was only hunting and gathering food and sleeping at night changed. But the body was designed by nature for a different task! So, humans now started working day and night, had to produce for others, work in a completely different manner, and over and above, the eyes were being used not for distant vision but more for close vision and that too at night along with the day and for longer durations. The structures of the eyes were not "designed" for that purpose!

A microscope and binoculars have two different purposes. The microscope helps in seeing very small objects which as such are not visible to the naked eye. Binoculars help in seeing objects very far which the eyes cannot see. Now imagine you have a microscope with you. You decide to see far off objects with the microscope, can you? Or imagine you have only a pair of binoculars and wish to see your blood cells, can you? In both the cases, the answer is no, because the two equipments have been designed for two different purposes and if you change their purposes, they would not function or rather malfunction! This is exactly what happened with the eyes:, they were being used for purposes for which they were not designed!

With passage of time, we see improvements all around the globe. These improvements are for the benefit of mankind and cannot be ignored. The human eyes are also intelligent and can tolerate if you make them work at night, for long hours in front of the screen, just like your mother who tolerates all your stupidity at home! But if you keep on misusing the eyes, then there are chances of error creeping in, just like how your mom gets angry one day and scolds you. The role of visual ergonomics at this juncture is how to ensure that the eyes as a window to the different visual information around us function in the most optimal manner without error or without discomfort and fatigue. This is important because Mother Nature has already designed the eyes and is not going to redesign them once again!

1.7 COMPONENTS OF VISUAL ERGONOMICS

Visual ergonomics as you have got a flavour now comprises many components. The most important components are the eyes, followed by the context or the environment in which users see anything. Then comes the age of the users followed by whether the user is sitting or standing or whether the user has to read or go through a set of information while remaining stationary or while on the move … the list goes on. Looks very confusing, right? All right, I will tell you a story which would make this a little clear.

Mr Benju is a 60 year old man. He is travelling to El Dorado and needs to board the aircraft from "Funny International Airport" which is about 40 kilometres from his residence. The flight is at 6am in the morning. Mr Benju tries to book a cab on his smartphone. As he opens the application, he cannot "read" the menus nor can he "understand" the different icons as they are too small. He takes the help of his nephew to book a cab. The cab arrives on time and he enters the same for the airport. At the airport, the driver asks him to pay the taxi fare. Mr Benju finds it "too dark" to identify the currency notes and asks the driver to switch on the lights inside the car. The driver is a little surprised as there is enough daylight and possibly an artificial light is not needed. Mr Benju gets down from the car and looks at the display board to know whether his flight to El Dorado is on time. The "font size" of the display is so small and there are so many flights and timings that Mr Benju is thoroughly confused. He has to ask a staff to help him out.

The story above narrates that when you present information to your user through the eyes, then there are a number of factors to be considered. These we call the component of visual ergonomics as mentioned in the first paragraph. The age of the user affects how well one can read and understand the information. Understanding the icons means you have to decode coded information. If the user is not familiar with an icon, then it's meaningless. The illumination inside the cab was too low for Mr Benju as he was an old man thus making it difficult to see the currency notes. The fonts on the digital display were not identifiable, and the user was mixing up different lines and thus had to take help.

1.8 WHERE DO WE GO FROM HERE?

Now we have some idea about what exactly is visual ergonomics. The next question is when we can apply it in design. In any field of design ranging from products of everyday use like a bottle full of oil in your kitchen to the controls and display inside an aircraft, it's all information being presented to the user mainly through the eyes. Thus, the font size, use of pictograms or icons, background and foreground colours, the ambient illumination, and a host of other factors determine whether the user will be able to see and understand what you are trying to present to them. This is visual ergonomics, trying to make visual information easy for you to see and understand fast so that you can take a proper decision and act accordingly.

This is followed by our next question, where could we possibly apply it? You can apply visual ergonomic principles practically on any product (for conveying product related information or how to use the product), space (helping people navigate in space by giving directions, names of roads, providing maps, showing path for cycling and pedestrians, etc.), and also on devices for aiding in communication between the user and the product (your mobile phone and the icons used for communicating to you different information).

The last question is why you should use visual ergonomics. The answer to this is, if I do not use visual ergonomics, then the information that I am trying to present to my user group will not be conveyed and my users would have problems in communicating with the design and would not be able to reap the maximum benefit out of it in the short and long run as well. Visual ergonomics is like your friend telling you all about the who, what, when, and how of the design through your eyes so that your design succeeds in the market and the users are happy.

Let me tell you another story of my grandfather. He was old when I was born but he was very prudent. Whenever I used to travel, he used to give me directions on how to reach that place. Along with the directions, he used to tell me about the different types of landmarks of the particular place. Listen, he used to say,

> if you are going to Eldorado, you will find a big fountain at the cross roads. If you turn right from the fountain, you will see a garment shop which is four storied. Just next to the shop you will get a very good coffee shop selling the best coffee that I have ever had in my life.

He used to repeat these time and again and it so happened that I had developed a visual imaginary (mental model is the scientific term) of the place and had no problem in navigating through the place when I actually went there. The other good thing that he did was to repeat what he said time and again. This "reconfirmation" on his part made the mental imaginary of the place absolutely crystal clear to me.

1.9 CAN WE TRUST THEM?

A common question that we often have to encounter is can we trust our eyes? Are they reliable when it comes to capturing information from the outside world? Let us draw an analogy to address this question. You have a dish television at home. The dish or the antenna captures the signals from the outside world and then feeds them to the set top box where it's decoded and converted into a different mode of signal which we view. Now if you ask, is the dish reliable in capturing information from the outside world? Yes, if the sky is clear and there is no rain or cloud cover. In the presence of rain or cloud cover, your favourite channel on the television screen shows "no signal". Does this mean the dish antenna is unreliable? Certainly not, because it becomes defunct when the weather conditions become harsh (rainy or cloudy), but when weather conditions are good, it works pretty well. The same could be said about our eyes. They are reliable and do not fail until you misuse them. For example if you burden your eyes and make them work in front of the computer screen for 12 hours a day without break, they're bound to malfunction. You will see everything blurred. Now for that you cannot blame the eyes, because they have been designed for distance vision and not for prolonged near vision.

That means our senses could be trusted keeping in mind their capacity and limitations and provided we use them prudently. If you use your screwdriver for chipping wood from wooden blocks, then it's obvious that the head of the screwdriver will become blunt and after some time it will fail to screw or unscrew. You go to the doctor if you have a health problem and you visit the barber for getting a haircut. Have you ever done the other way round? You know very well what's going to happen then. Thus, don't fool around with your senses and they will do their work most reliably.

1.10 THE WHY AND HOW OF INFORMATION

We are bombarded with information every day, and at times, these are in huge numbers. The eyes are selective and they do not permit all the information to enter and thus overload themselves. Thus, your eyes are sensitive and work like gates permitting what is necessary and filtering out what is not necessary. When it comes to any information and visual information is no exception, there are a few aspects where we need to be a little vigilant from the user's perspective. The first thing that we need to ask ourselves is, why is this information important? Can we not do without this information? Is there any other way of presenting this information?

Once you are sure that you need to feed the information to the users then the next question is, in what format? Is it necessary to present the day temperature to a common person like 23 degree Celsius, or should it be presented as

23.4456 degree Celsius? The user group would indicate how the information is to be presented so that it's encoded and understood very easily. In majority of cases, it's the common people to whom we need to present the information so as to make their life comfortable and easy.

1.11 WHERE THINGS GO WRONG

Life is always not comfortable. You might not be feeling well every day. There are instances when the context might not be conducive for presenting the information. If you are moving in a dark lane, you would not be able to read the name of the road because of the ambient light or because of the manner in which the name of the lane is written. If you have to compare the performance of boys and girls in examinations, then what could be the best way of depicting it? The bar diagram could possibly be the best way of comparing two sets of data especially for the common people. An elderly user suffering from cataract would not be able to differentiate between many colours or read warning labels on packaging clearly. Thus, in such cases, the context and the people are also important and it becomes very challenging to present the visual information to your target users. Just try operating your smartphone in bright sunlight! The glare doesn't allow you to see anything. That does not mean that the icons have not been well designed.

So, we can see when everything is in order, the users can see and understand the information presented to them (at least what they are supposed to see). The problem arises when users are elderly, have visual defects, the context of use or the environment around is such that the users are unable to see the visual information properly and things start deteriorating. This is when visual ergonomic principles can help to a great extent. In the previous paragraph, such adverse conditions have been depicted and this is where the visual designer feels handicapped and thus requires the prudent application of visual ergonomic principles.

1.12 ROLE PLAYED BY VISUAL ERGONOMICS

Visual ergonomics is like the school principal. She/he heads the school, looks after the academic and administrative matter, and is vigilant about the welfare and education of every child in the school. Visual ergonomics ensure that all the pieces of visual information are presented in a manner that the target users are able to understand them and this further helps them to take a prudent decision. For example what family of typography would be better understood by the target users when some information needs to be written on a milk carton is best decided by visual ergonomics. If there are red and green colour-blind users, how to prudently use different shades of red and green on different products, spaces, and communication devices is what visual ergonomics dictates. If you have to read the road signage from a distance of 12 feet, then what should be the height of the typography is also dictated by visual ergonomics. Thus, in a nutshell, visual ergonomics ensure that visual information is made usable and much more human to the target users.

1.13 APPLICATION AREAS

Wherever there is visual information in the form of text and images one can apply the principles of visual ergonomics. If you want to publish a newspaper, you can use the principles of visual ergonomics to decide the layout, font size and spacing and colour of the text, placement of visuals, etc. in a manner that users are able to read it with ease, get maximum satisfaction, and read the content with speed. If it's a safety symbol for slippery road for motorists, visual ergonomics tell you where exactly you should place the symbol, what should be the size of the symbol, and the preferred colour of the same. In other words, visual ergonomics can be applied in any domain where there are users and where one needs to communicate some information to the users through the eyes. We all know, through the eyes the information very easily enters the mind and the "soul". If we look into different sectors, visual ergonomics can be applied with equal importance in all of them. It can be applied in the domain of healthcare when it comes to designing of different displays, medicinal packaging. It can be applied in the entertainment sector like cinema and stage performances where the principles could be applied for deciding the typography style, typography font, how the casting should come. On the stage, visual ergonomics takes care of the colour of light and other visual elements used to create special effects. In transportation, they play an important role in designing traffic symbols and directional signage.

1.14 DESIGNING THAT WOW

Today visual ergonomics is used for generating the experience of users when they interact with a product or enter a space. When you entered that small coffee shop for the first time with your friends you felt, "wow what a beautiful place". Nice graphics all around, some favourite character of yours drawn on the wall, those nice quotations on the walls … all added to that "feel good" factor in you and that's why though the coffee might not be that great (in fact your mom makes better coffee at home!) you liked the place and revisited with your friends whenever you wanted to relax. So visual ergonomics convey information through the eyes and generate in users that feeling of "wow" and breathe life in the space. When you picked up that shoe in the shop, you just liked it not because it was only comfortable but because of the graphics and the colour combination and finally your dad had to buy those pair of shoes which were pretty expensive!

1.15 COMPUTER WORKSTATION

Visual ergonomics also contributes to how to design your computer workstation so that you do not suffer from low back pain and have that blurred vision. In this book we are not going to talk about that in much detail and will just touch upon it superficially.

1.16 KEY POINTS

A. Eyes are the most important windows in the body that help the users in communicating with the world outside.
B. The different senses in the body at times work in isolation and at times work better in pairs.
C. The eyes are the most precious gift of nature.
D. Visual ergonomics augment the communication between users and the outside world through the eyes.
E. Eyes are meant and designed for distant vision.
F. Visual ergonomics work in a context and are also dependent on the user group.
G. Our sensory channels are extremely reliable provided we use them prudently.
H. Visual ergonomics can be applied in all the sectors serious and non-serious both.
I. Visual ergonomics help in creating that experiential component in space and products.

1.17 PRACTICE SESSION

1. Pick up a soft drink bottle and try to list down the different visual ergonomic issues in points of communication between the information on the bottle and the eyes.
2. When you go to the market, list down the communication elements you encounter on the road and how many of them you can understand.
3. Pick up any product of your choice from the market and try to identify the different visual ergonomic components in them.
4. Select a space of your choice and list down what visual ergonomic elements you could incorporate in the space to make it livelier.

The Coding of Visual Information

2

Overview

This chapter discusses the different visual ergonomic principles in coding information. This coding of information could be in the form of icons, symbols, or graphics. The visual ergonomic principles that should be kept in mind while designing such coded information will be discussed in this chapter. In this chapter, the different visual ergonomic principles and their application in designing textual materials and usage of typography are discussed. The chapter touches upon different visual ergonomic aspects of information design for different products.

2.1 THE WORLD OF ICONS, PICTOGRAMS, AND SYMBOLS

An elderly person (Mr Leto, aged 68) living in a village in a remote rural area had to visit a super speciality hospital in a town. The person Mr Leto could not study during his childhood as his parents were very poor. He made both ends meet by cultivating the small piece of land he had inherited from his father. Though Mr Leto could not read or write, but he still decided to go to a big town for visiting the hospital as he wanted to sort out his health problems which the doctors in rural areas could not diagnose. He boarded the bus and this was the first time he was going to a big town in his life. It was a four hour journey with small breaks in between from his house to the town. As the bus started, he started looking all around and enjoying the beautiful landscape. As the bus entered the town, he could see only buildings and a lot of things

DOI: 10.1201/9781003369516-2

written here and there and could not understand anything. He could see on the roadside a "drawing" of a male and female and that too an outline. This he could see all through the town but had no clue. The person sitting next to him was engrossed in using his smartphone. Mr Leto never owned any smartphone or even a mobile phone. He looked at the person and saw him putting his finger on different images of different types and colours. Mr Leto was just wondering how the person could understand what those "drawings" of wheel, people, clock, and loads of other things actually meant. He looked outside and found that the bus was taking a left turn and an arrow placed on a pole pointed towards the direction the bus was moving. Finally, the bus arrived at the hospital and since he had requested the young man to inform him when the bus arrived at the hospital, he had no problem. Mr Leto thanked the young man and got down from the bus.

Now the big challenge started for Mr Leto. He stared at the main gate of the hospital and had no clue where he should go. He asked the security at the main gate who asked him to go to the reception. Mr Leto was now lost; "where is the reception?" he asked the security. Go straight and then turn right and look for the buildings and one of them is the reception and you find it written on the building. Mr Leto did so, but he was not literate and had no clue which building was reception because every building had scribblings on it and looked similar.

In a world full of uncertainty, users like Mr Leto and others need hand holding so that life becomes easy. You cannot always depend on others in life. Icons, pictograms, and symbols reduce uncertainty for the user in different domains. Visual ergonomics can play an important role in their design and make them much more acceptable to the target users. Mr Leto saw the young man sitting next to him in the bus put his finger on different "images" of wheel, people, clock, etc. which were examples of "icons" (Figure 2.1).

These store information in coded form and have resemblance with the real world, which is it in reality exists. When the bus was taking a right turn, Mr Leto observed an arrow pointing towards the right on the road (Figure 2.2); this is an example of a pictogram, which is a communication modality again in coded form to indicate direction of movement. The arrow is communicating a small word or phrase "turn right".

As the bus entered the town, Mr Leto observed that outlines of male and female drawings were found in a lot of places in small buildings which looked like toilets. This is an example of a symbol, which is again coded information, but for first timers, it might not be clear. It becomes clear only when users are exposed to them repeatedly, and only then do they get to learn that these are for toilets and the male and the female symbol indicates male and female toilets, respectively (Figure 2.3).

These elements dominate the world around us and feed us all with information of different kinds. We see them in space, products, and other areas

FIGURE 2.1 Icons on a mobile phone. Photo by ready made: https://www.pexels.com/photo/crop-faceless-person-showing-application-icon-on-smartphone-screen-3850274/.

as well, and they help in conveying to us how, where, what manner, and all information related to these in our day to day life. Visual ergonomics play a role in making them much more "humane" or user friendly so that all the users irrespective of their background are able to understand them. Here, Mr Leto could not understand many of them, as the visuals were not designed in tandem with ergonomic principles.

2.2 APPROACHING VISUAL ERGONOMIC APPLICATIONS IN ICONS

When you were a kid and used to go to different places (Figure 2.4), you might have noticed that people used to ask you what the names of your parents are.

This was the case only when you were alone or in the company of other adult members or with children of your age. But the scenario was different when you were with either of your parents. Did you notice that no one ever

FIGURE 2.2 Pictogram of direction on a road. Photo by Maria Orlova: https://www.pexels.com/photo/corner-of-aged-house-near-direction-post-in-town-4915846/.

asked you this question of who you are? This is what happens with icons. When they are in a family or in a context, they are easy to identify, but the moment they are alone and in a different context they lose their identity. When you see the icon of "file" either on your smart phone or laptop then you can immediately identify it. If I show you the icon of the same file on a piece

FIGURE 2.3 Symbol of a toilet. Photo by billow926: https://www.pexels.com/photo/sticker-on-tile-wall-5338294/.

of paper or on a separate page in a digital medium then it loses its meaning. Thus, the first lesson learnt is that while designing icons, please look into the context of usage. You also need to look at the other icons which are to be located in the vicinity of the same and treat all icons as a "complete family" so that each one is understood very easily by the user. Just like you are identified in the presence of your parents, similarly at your friend's house, when

FIGURE 2.4 Your parents with you. Photo by Emma Bauso: https://www.pexels.com/photo/family-of-four-walking-at-the-street-2253879/.

you go with your parents, your parents are identified better by your presence as people say oh, you are the parents of Beatriz (your name). So when you design an icon, make sure you design the entire set of icons as well if possible or design them in tandem with the other icons so that they "gel well" in the family of icons.

The next issue is whether icons should be coloured or black and white. Our eyes are designed in a manner that they can see colours, and a colour at times helps the users to identify the icons better compared to the same in black and white. So we suggest using coloured icons if cost is not a factor. Now this statement is a general statement and cannot be used everywhere and under all circumstances. There are pure red and pure green colour-blind users who would perceive the red and green colours in shades of grey and hence the colours pure red and pure green would be a problem for them.

What about details of the icon? Just imagine you have ten seconds to look at an icon and identify its function at any multimedia interface. If the icon in such a situation is very detailed then there could be two problems; one it would take time to decode the icons, and second, if the size of the icon is reduced or the icon is small then all the details would get hidden. This is because the eye would take time to scan the details of the icons. Now just imagine. Your brother is lost and your father asks you to go to the main road and locate him and bring him back. You go to the main road and are confused; there are a

FIGURE 2.5 Trying to locate your brother on a crowded road. Photo by Hert Niks: https://www.pexels.com/photo/busy-street-3224225/.

large number of vehicles and people moving down the road. You find it very hard to locate your brother (Figure 2.5).

If the road had been empty, it would have been relatively much easier for you to locate your brother. Let us take another example. I give you a white paper wherein eight oranges are drawn and I ask you to tell me how many oranges are there (Figure 2.6). It is very easy for you to identify eight oranges

FIGURE 2.6 Trying to figure out apples from a group of fruits. Photo by Ian Turnell: https://www.pexels.com/photo/assorted-fruits-stall-709567/.

against the white background of the paper. Now I hand you papers on which there are oranges, apples, pears, bananas, and pineapples are drawn in different numbers. You are now asked to count and identify the number of oranges. Now the task becomes difficult for you. You have to identify the oranges from the crowd of other fruits and thus you take much more time and might even make mistakes in counting the oranges. Thus, it's preferred that details of icons are kept at bare minimum.

What about the size of the icon? When you develop the image of any object from the real world, the size does not matter as long as you are able to recognise it as a scaled down or miniature form of the real world. For example when you see that familiar image of the trash can on your desktop computer, you recognise it at once, because it resembles the real world waste paper basket, and is about 20 to 30 times smaller than the real world waste paper basket. As it retains all the features of the waste paper basket you have no problem identifying it. If the size were to shrink further and the details get lost? If the top and the bottom of the icon were not distinguishable, then identification of the icon would be difficult. Thus, the size of the icon retaining all the characteristic design features of the real world object is not a problem. But the moment the features or identifying features are lost, then comprehension of icons is lost. You are looking at an icon of a person. Now if in the icon you find that the head and the legs are missing then can you identify it as a human? No, it is not possible.

Should we reinforce text with icons? Ideally no, because in that case the purpose of an icon gets lost. But on the other hand, there are circumstances where your users need reconfirmation especially in critical contexts. You know that one of the human characteristics (refer to *Ergonomics for the Layman: Application in Design* by the same author for more detail insight into it) is humans expect reconfirmation from a system. If a doctor is operating a medical equipment (which is a serious domain, you stand between life and death of a patient), she/he has to come across many icons at the device interfaces. Some of these icons are very critical because they are related to administering a lifesaving drug. Now the doctor attending to a serious patient is already under stress and cannot afford to look for the meaning of an icon. In such a context, a textual reinforcement helps the user. When the icon is clicked and the name of the function pops up as textual information, this acts as a reconfirmation to the user that he/she is operating the desired control. But apart from this, providing textual information helps elderly users who require reconfirmation as they are not able to remember what each icon stands for.

All around us what we see are static icons or icons which are stationary. What about dynamic icons? Icons which exhibit movements? Should we use them at different interfaces? The answer to this is not that simple. Humans register dynamic information better compared to static information. But saying that, if everything moves in front of your eyes, then it becomes very distracting and might lead to error. Moving icons are used to depict certain movements at the system interface. It could be used to depict "file being copied", "file being downloaded", "engine overheating", etc. which requires the attention of the user (Figure 2.7). In these contexts, animated or dynamic icons work very well.

FIGURE 2.7 Dynamic icon of a file being copied. Photo by Cup of Couple: https://www.pexels.com/photo/crop-person-with-laptop-and-coffee-6177618/.

2.3 APPROACHING VISUAL ERGONOMIC APPLICATIONS IN PICTOGRAMS

The word "pictogram" as the name suggests emanates from "picture" or mental picture in users. The pictogram is a pictorial depiction or a drawing which as such does not have any resemblance with the world around you like the icons. These are used to convey to the users a particular message, without writing it down in a textual form. For example as you walk down the street you might come across different types of arrows pointing towards the right or left and also sometimes towards the sky! The arrow towards the sky indicates (Figure 2.8) that you have to move ahead in front and this is not "digested" well as we have to decode it as forward movement. So the arrows though they do not resemble anything do convey "move right", "move left", or "move straight". While designing these pictograms you need to take care that they "pop out" from the surroundings and make their presence felt to the user as if saying "I am here to help you". The placement of these pictograms should be within your cone of vision so that they catch your attention. If you are walking down the road and cannot see the person selling newspapers then you cannot buy them. Similarly, when you need direction whether to move right or left, you need help and it should be in the "expected area" of the visual field. This expected area is where users would look for help. Sometimes we need to find out where users look for help and place pictograms at that position. The size of the pictograms is again dictated by the distance from where you expect users to see them. In the later chapters we are going to talk about them in much detail. Normally, do not complicate pictograms by adding too much detail because the users might get confused.

2.4 APPROACHING VISUAL ERGONOMIC APPLICATIONS IN SYMBOLS

This is essentially a combination of a visual graphic and a pictogram and is very strongly related to the population and the way users expect things to happen. Symbols are very much official and formal in nature, just like wearing a suit and tie to office. This combination of graphic and symbol in the initial stages might not be understood by the users at the first go. Users have to learn their meanings and then they are able to recall them.

FIGURE 2.8 Arrow pointing towards the sky. Photo by Christopher Flaten: https://www.pexels.com/photo/bright-road-sign-with-curved-arrow-on-lawn-4266243/.

The context of use of these symbols plays an important role in their easy decoding. You see male and female images on toilets. Those images in isolation or for the young users might not make sense, but as the novice users get exposed to them or rather repeatedly get to see them, they learn that these are toilets for male and female users. Placement of these symbols and their size are very important attributes in ensuring that they are designed in tandem with ergonomic principles.

2.5 THE THREE IMPORTANT VISUAL ERGONOMIC ISSUES IN TEXT

If a person is literate, then reading a text is not a problem at all. This text could be on a product, a roadside advertisement. If it's a dimly lit room, or if it's dark outdoors, then reading, decoding, and understanding any textual information becomes difficult for all of us. Pick up a strip of paracetamol tablet (used for controlling fever). Switch off the lights in your room and try reading the expiry date of the medicine. You had it! Very difficult or almost impossible to do that. Now turn on the lights, it's now relatively easy. Ask your grandfather to read the expiry date on the medicine strip. First, he would put on his spectacles (for close viewing) and then after failing to read would ask for a hand lens with which he would try to do the task given by his grandchild.

The context depicted above indicates that when everything in life is "hunky dory" you need not worry, reading textual materials is a "cake walk" for the users. But if the context changes, like the illumination level is low, it's dark, or your users are old or have problems with their eyes, it's necessary that we resort to some visual ergonomic attributes that would help us sort out and address these problems. For handling any problem in this world, we need to know three things: what is the type of problem, where is the location of the problem, and lastly, how much is the problem. When you read a text, there are three things associated with it from the user's perspective.

First is that the text should lie within your cone of vision. If you cannot see it, or if it's not within your visual cone, then there is no question of reading it. If there is an advertisement for the availability of hotel rooms written on a board and on the road mounted on a one thousand feet tall post, can you see and read it from your car? Definitely not, because it's not at your eye level and in your visual cone. Thus, any information written in textual format should be at your eye level and within your visual cone.

The second attribute is that the textual material which is written needs to make sense. I mean the text should clearly convey to you all the letters in their true form. That is, the *A* should look like *A*, *O* should look like a *O* and not be confused with a zero, *I* should look like an *I* and not be confused with one.

The third attribute is that the users should be able to read the textual material with ease. This is where one should factor in the fact that the distance between the lines and that between the letters should be such that readers have no problem in reading the text.

So, for any text to be read by the users easily, within a short span of time, and correctly, one has to factor in the above three. If these three criteria are not fulfilled, then the textual material is of no use to the users as they cannot use it or no information can be conveyed to the users.

2.6 VISUAL ERGONOMICS OF TYPOGRAPHY

We are all familiar with text and textual information conveyed to us on different products, spaces, media, and roads. These textual materials comprise letters, numerals, and other signs of the English language (that's what we are discussing in this book) or any other languages over the globe. These letters and numerals are of different styles, sizes, spacings, etc. and play a very important role in communicating the information to the users. The manner in which this jugglery of different elements is done with the letters and numerals dictates at times how effectively the information is conveyed to the target users.

You are writing an application for a job for the first time in your life. If your handwriting is good and appealing, then irrespective of the content of the letter, the very style of your writing would appeal to your employer when he just looks at the write up even without going in detail. This is called the first look of the textual material. Here the arrangement of each letter, style, colour, and similar elements plays a very important role in conveying the information to the target users in the most effective manner. These arrangements are extremely important in differentiating textual elements as to serious, warning, danger, or just casual. The letters and numerals convey different types of messages when used in isolation and when used in different combinations. That's why the letters and numerals can also be referred to as elements in the textual space.

In general, the different typographic variations mentioned above should adhere to the visual ergonomic attributes of having it within the visual cone, understanding clearly each individual letter and numeral as they are, and be able to read the information ensuring the letters and numerals are spaced and arranged in the space properly. The challenge is that there are certain circumstances when the textual material might not be understood by the target users. So what are these unfavourable circumstances when users fail to understand what is being communicated to them? The following paragraph explains that in detail:

Incidence one: You are driving your car and it's already dark. There is suddenly a power outrage and all the lights on the road go off. You are looking for a specific shop which sells sea food, but the darkness around doesn't allow you to see or read anything clearly. Adding to this worst situation, it starts raining and it is raining so heavily that you now cannot see anything even two feet away. These are "unfavourable circumstances" where communication of textual information is difficult unless you apply visual ergonomic principles.

Incidence two: Mr Raju is driving his car slowly on a Sunday morning and sees a shoe shop giving sales on all the shoes. He tries to look at the price of the shoes displayed on the showcase from his car. Unfortunately, the text is so small that he is unable to read the price from his car seat. He is disappointed and drives back home. This is a case where the "distance" from where the user tries to decode the textual information is too far and thus the communication becomes difficult.

Incidence three: Ms Nani is 80 years old and suffering from low vision. Recently she has been detected with an advanced stage of cataract in both eyes. One day she was having a severe tooth ache and had to go to the clinic of the dental surgeon all alone for a check-up. The receptionist at the clinic handed her a form that she needed to fill up. Ms Nani is unable to read anything as everything looks blurred to her. She finally asks a person sitting next to her to help her out. This is a case where the user Ms Nani had "impaired vision" and thus communicating with textual information became very difficult.

Incidence four: You have gone to a supermarket and you notice that the place is literally bombarded with textual information all around. But when it comes to COVID related information and the precaution to be taken, it's written in a manner which is different from that of other textual information and "pops up" from the rest of the information in the vicinity. This type of information is known as important or "critical information".

All the four incidences above indicate that letters and numerals used to communicate information under normal circumstances generally follow the principles of typography and hence communication of the desired information is not a problem. But when they don't, like the incidences mentioned above, then one needs to go for visual ergonomic intervention to ensure that they do communicate what they are supposed to. This is a big challenge for visual designers.

2.7 DIFFERENT ASPECTS OF TYPOGRAPHY

The sizes of typography characters are very important and one needs to calculate that. The height to width ratio is an important parameter when it comes to deciding the typography size. For the letters, it's referred to as the "stroke" height to width ratio. This stroke height to width ratio actually changes and is not constant. It's dependent upon ambient illumination, and under optimal illumination and depending upon the background and foreground colour, it changes. Under very low levels of illumination, thick letters have been seen to perform better. Dark lanes, dark corners of room, or products on which textual materials are to be read in dark should have thicker letters compared to normal ones. It is believed that if the illumination level is very low and you have low contrast on the product that you have the information on, then possibly it's a good idea to reduce the width to height ratio.

2.8 JUGGLERY WITH TYPE AND STYLES

You have a lot of friends, but is every one your best friend? That cannot happen. You will always have preference for some friends and less preference for others. Have you ever thought why all your friends in your class are not best friends? In reality this cannot happen. It takes time to build relationship, and a strong relationship happens with those who understand and accept our characteristics and we also accept theirs. In the world of typography "font" is a popular term used to denote "style" of the letters and numerals. These styles of the letters play a very important role in communicating information. If we know the preferred style of the users, the textual information is read with speed and ease with minimal error by the users.

We all live in a family. There are ups and downs in a family. You do not perform well in the examinations and you decide that you would leave the house forever because your parents have scolded you. Now your parents are in a fix. You are not going to listen to them. So what do they do? They call your best friend and ask him to convince you to reconsider your decision and stay back. The trick works, because you are "communicated an information through your best friend". In typography also users have some preferred font family, and when information is conveyed through them, they are effective. In fact, the different font families also indicate to the user the seriousness of the information being conveyed. Some fonts are believed to be serious and some funny. When users like a font family, they develop an image of that family and later on that image engraved in the brain helps in decoding the information very effectively.

With all font families, normally there are two styles available for conveying information. These styles are differentiated on the basis of particular characteristics. Mr Leto and Mr Peto are two brothers and look alike. The

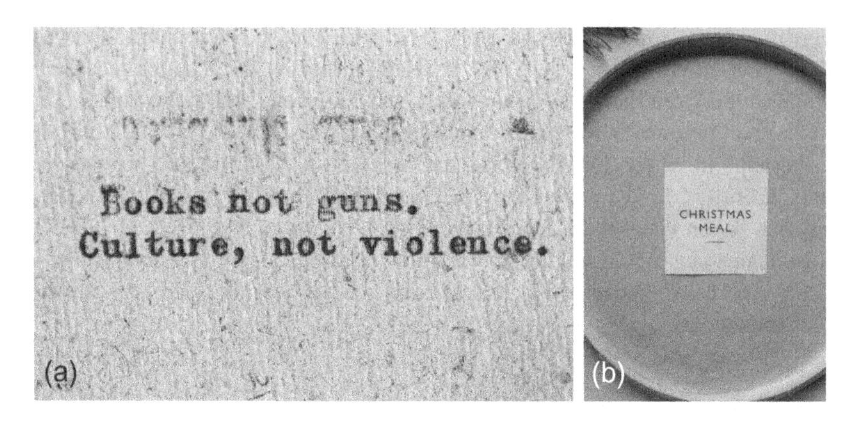

FIGURE 2.9 Serif and sans-serif font. (a) Photo by Karolina Grabowska: https://www.pexels.com/photo/anti-war-sentence-on-old-paper-5993566/(SERIF). (b) Photo by Monstera: https://www.pexels.com/photo/tray-with-greeting-card-on-table-5709029/.

only feature which helps to differentiate between the two is that Mr Leto has a moustache with curly ends and Mr Peto has a moustache with blunt ends as if someone has chopped it off at the end (Figure 2.9). In the font family also there are two styles available. One has a moustache or extension at the end of the letters called serif and the other has the ends chopped off and is known as sans-serif. The serif or moustache font is a little stylish and in print medium looks pretty well. Sans-serif on the other hand looks very "polite" and is used mainly in digital medium because it causes less blurring because of the blunt end. The moustache font generates a better or more defined pattern compared to the non-moustache one and thus its reading ease and speed is more. Our eyes are very good in reading patterns and this is explained in the next paragraph.

2.9 CASE IN INFORMATION

If textual information is presented to you in all capital letters and upper and lower case, which piece of information can you read fast and with minimal error? Confused right? Well, if the information is long and is more than three to four words, then upper and lower case is better. This is because when you write in upper and lower case, there is a pattern generated around the words. The eye actually follows the pattern and does not read each alphabet and thus you read it fast and with ease. But when I need to shout or shout aloud to draw your attention like STOP, GO, DANGER, the all caps is better because it stands apart from the crowd of text and says "I am here" and follow this.

When two words which are very confusing are used side by side, then all caps helps in better differentiation and tells you to "examine it very closely and carefully". For example "fry "and "dry" might be confused as they sound similar. When you write the same words in all capital like "FRY" and "DRY", it indicates close examination and warns you to be careful as it stands apart from the crowd. As if two policemen in a crowd warning you "beware of pickpockets".

2.10 READ WITHOUT EFFORT AND WITH PLEASURE

If you follow some small tricks in visual ergonomics, you can ensure that your text can be read easily and users would enjoy while reading the same. This is where the spacing between the words plays a very important role. Uniform spacing between words helps the brain develop a "mental model" or set pattern and thus the reading speed and ease both increase. Let's take an example. Suppose you have been asked to pass a ball between two sticks, ensuring the ball doesn't touch the sticks. The sticks are uniformly placed, that is the distance between the sticks is the same and there are 20 such sticks. Initially you are careful and a little slow but gradually you learn the technique (we have a term for this called "mental model") and thus later on you do it faster and easily. Now if someone changes the distance between the sticks to be unequal, that is in some cases the distance between sticks is more, and in some cases it's a little less and all the distances between the sticks are not uniform. What happens now? You have to pass the ball carefully for all the sets of sticks and hence you become slow because sometimes it's more and sometimes it's less (Figure 2.10).

This is what happens when you justify your text. Just to make the text look good by having straight edges on both sides the spacings between the words become non-uniform and the eyes get stuck at the unequal spacings and just wonder "what the hell is happening" and it slows down your reading speed and ease. Whereas if it's left or right aligned, the spacing between the words is uniform and your eyes develop a typical pattern like it happened when you passed the ball between two sticks when all the sticks were uniformly placed. Reading becomes easy and fast.

FIGURE 2.10 The stick and the ball experiment. Photo by max laurell: https://www.pexels.com/photo/sticks-over-plank-on-water-6251784/.

2.11 BE CAREFUL WHAT WORD YOU SELECT

When you were a child and did a lot of mischief, your parents used to tell you, "don't do this". You did not like this statement because it was harsh to you because of the "don't" in it. Rather when your parents after your mischief told you, "son you should do this" that sounded much better to you because it had a "direction" in it and the former was "direction less". In visual design when playing with textual information you need to follow this so as to win the hearts of your users. Always give users a direction rather than keeping them directionless.

2.12 KEY POINTS

A. Icons, pictograms, and symbols convey information in coded form in tandem with users' mental models.
B. Icons resemble the real world, pictograms are abstract and communicate information, symbols are a combination of graphics and pictograms which have been accepted by users.
C. Visual ergonomic aspects of placement, visual cone, size, colours, and details need to be factored in while designing them.
D. Visual ergonomics of text should focus on visual cone, identification of letters and numerals, and proper spacing between letters and numerals.
E. The change in context and the user demands a special approach in visual ergonomics of textual information.
F. Always use the preferred font family of users to ensure ease of reading.
G. Serif fonts are mainly to be used in print and sans-serif in digital medium.
H. Left or right alignment of text enhances reading speed and ease.
I. Upper and lower case generates a pattern and hence in long text use them for ease of reading.
J. All caps emphasise important actions.
K. Use words and sentences that give users a direction of what to do.

2.13 PRACTICE SESSION

A. Pick up any multimedia device and try to list the icons which are not easy to interpret. Try listing the visual ergonomic attributes in them.
B. Pick up any pictogram from the road and try to do value addition to it by adding visual ergonomic issues.

C. Pick a traffic symbol like "no right turn" and try to evaluate it in the local context. List down the visual ergonomic issues.
D. Pick up a text from any pesticide bottle and apply visual ergonomic principles in designing the same.
E. Design the textual information on a cough syrup bottle from the visual ergonomic perspective for your elderly grandmother.

Typography and Colour

3

Overview

In this chapter the visual ergonomic aspects of colour and its usage are discussed. The role of colour in typography and other visual elements in textual and visual information design is discussed in length. The chapter starts with different attributes of colour in visual design and then gradually moves to different areas of application and to establishing the relationship of colour with typography. This chapter introduces to the readers the different ways and means of arranging text and typographic elements in tandem with ergonomic principles. The different principles are discussed, and lastly, how to use each in the designing of textual information is enumerated in detail.

3.1 COLOUR AND ERGONOMICS

As discussed in the previous chapters, our eyes are "designed" for seeing everything in colour. So, colour does make the world around us "colourful"! But provided the other factors are kept constant. That means if you are presenting information to the users in colour, in general they are better understood compared to if the same is presented in black and white. Just imagine you are reading your favourite cartoon comic in black and white. If someone offers you to read the same in colour, you would immediately opt for the coloured cartoon comic. You are shown the world map in black and white. Then you are presented with the same world map in colour. Your understandings of the information presented on the map is much better in the coloured map compared to the black and white map. So, the use of colour in conveying information through the eyes is definitely much more effective compared to not using them. But be very careful! This is a very "general statement" and always not true. You will get to know in the subsequent sections the reason for the same.

DOI: 10.1201/9781003369516-3

An elderly person (age 82 years) was sitting in a completely dark room in a tourist office and was presented with a road map. The road map was in colour and represented the route to different destinations in the city. "What is this?" The elderly person shouted. "I cannot read anything. Please make some arrangement for some light". The example here indicates that even if you present information to users in colour, the ambient illumination, the type of task (reading fine prints on the map for example), and the condition of your eyes (elderly persons have eye problems like a cataract, lessened accommodation power of the eyes, etc.) would not help. All these factors are to be taken care of at a macro level to ensure that you are able to convey the required information to your users. This proves that even if you use colours to represent information to users there are limitations for the same. Users have different preferences for colours for different objects. You might like a red coloured car but not necessary a red coloured trouser! Some colours also elicit different feelings. For example light blue is believed to give a sensation of calmness and makes a space look bigger than what it actually is. While you are using colour, you have to keep in mind three things as well: the light source, your user, and the information that you want to convey to your users.

3.2 TRYING TO QUANTIFY COLOUR

While trying to use colour in presenting information, you need to know a few aspects of it. Mr Rascal goes for a morning walk in the rose garden close to his house. The garden has roses of many different colours like red, yellow, pink, white, and so on. He walks and enjoys the colour and the fragrances of the roses. The red roses that he sees are in the purest form because it's the colour which has no additives and dominates (like some of your friends in the class who project themselves as "very important persons"). Mr Rascal (age 58) once purchased a trouser for his wife (age 55) for her birthday and brought it home. When he gifted his wife this trouser, his wife reacted by saying, "oh my God, it is so bright!" "My dear", she said, "at this age if I wear such a bright coloured trouser, then people are going to laugh at me. You should have brought a little dull coloured trouser for me". In this case, Mr Rascal's wife is talking about another aspect of colour which deals with its brightness. Mr Rascal's wife also added, "dear please get me a trouser which is red but a little dull red". So, for the laymen like us, three aspects of colour (there are many more) can be factored in namely dominance, brightness, and purity. This is to make your life simple at the beginning.

In visual design you are often tempted to only go with the "colour wheel". There is no harm in using the same but remember that the colour wheel helps you to anticipate (grossly) what would happen if you mix colours of different types. So, definitely you can use it as a tool for "playing around" with different colours, mixing them up, and coming up with some fanciful combinations. But when it comes to designing in serious domains such as imaging in medical sciences, then the colour wheel alone won't help and you need to augment the colour wheel with other techniques in visual ergonomics. Use of pure blue colour is to be done with caution because the central part of the retina (photographic plate of the eye) which is an important part in seeing objects is lacking in the cone cells (which help in detecting colour) which are capable of handling blue colour. Thus, eyes would perceive pure blue in the central part with a little blurry outline.

3.3 TYPOGRAPHY AND TEXT

You are often confused as to which would be a better choice, black text and a white background or white text and a black background. This could be extended to should we use dark text on a light background or light text on a dark background? Confusing, isn't it? This question is incomplete if someone asks you this. We have seen before that the ability of our eyes to see visual information is dependent to a large extent on the ambient illumination level. If the ambient illumination is low, the aperture in the eye (called the pupil) grows bigger to permit more light. The dark background comprising a larger dark surface area compared to the lighter text facilitates this. More light enters the eye and visibility enhances. If ambient illumination is normal, then a light background and dark text are preferred. In this case the pupil shrinks because light is there in the vicinity and white colour reflects light as well (Figure 3.1). Shrinking of the pupil narrows the aperture of the eye and also "cuts" unnecessary light rays and thus the user is able to see the textual material (print medium) much better.

3.4 IS COLOUR USAGE ALWAYS GOOD?

Though we use colour in representing visual information in many domains, it has to be done in a prudent manner. When it comes to producing very fine details, colour is not of much use and we resort to black and white in those

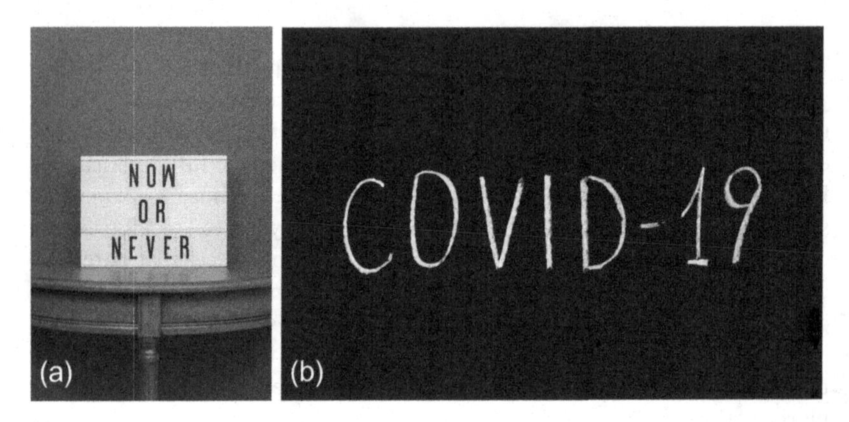

FIGURE 3.1 Black text on a white background and white text on a black background. (a) Photo by Alex Azabache: https://www.pexels.com/photo/white-letter-board-on-wooden-surface-4084146/(black text white background). (b) Photo by Anna Tarazevich: https://www.pexels.com/photo/covid-19-lettering-text-on-black-background-5697253/ (White text black background).

cases. In maps, it's always recommended that if you want your users to differentiate between different categories of information like different states in a country, height of different people, edible oil consumption by different races, then instead of using different colours (you only confuse people if there are 30 states with 30 different colours), in this case using the "dominance" of a colour in different degrees is a better option.

This dominance aspect of colour plays a very important role in interpreting data in medical imaging. If you have been to any hospital for an imaging, then you might have noticed that the radiologist is keenly studying the photographic plates which are often in colour. The doctor actually looks for abnormalities in the images. This is where instead of using different colours the "dominance" of a particular colour is used to differentiate between normal and abnormal components of an organ. Every organ in the human body exhibits some chemical reactions but they are not the same. This difference in chemical reactions helps in differentiating normal and abnormal cells by using the colour dominance in the images. The colour dominance in normal cells is different from abnormal cells. This is what the doctor first tries to identify and then goes into the details. Let's take an example of a soccer match between two teams. The teams are wearing deep blue (more dominant) and light blue (less dominant) jerseys. After the game starts, you will be able to see the players of two teams moving into each other's defence. If you now take a photograph of the players from above, you would see that on the field there are areas where a lot of players wearing dark blue jerseys are present and in some other areas players with light blue jerseys are present. If the team with the dark blue jersey dominates, then you will find that one area of the field is full of dark blue jerseys (dominance) and the other part with light blue jerseys. So, looking at the jersey colour on the field you can say which team is playing better than the other.

3.5 QUANTUM OF COLOUR USAGE

Our obsession with colour is so great that we tend to use it everywhere. Just try this yourself. On a big white sheet of paper, take some deep blue colour on a needle tip and make a tiny dot, and with another needle, take some red colour and make a tiny red dot. Now try to differentiate the dots created by the needle tip! Can you? What is the problem? The problem is, when you use a small amount of two colours, it's almost impossible for the eyes to differentiate between them. So, everything in the world is not colourful! Now try this experiment yourself. Draw very small triangles, squares, and circles. It should be very small and in three different colours. Ask your friend to write the names of the colour for the square, circle, and triangle. They will all fail to do so if the geometrical forms are very small (Figure 3.2). Why? The reason is that they are so small that in that small area, it's impossible to identify the exact colours. Thus, all is not colourful! When it comes to distinguishing small elements, geometrical shapes' colours don't help. It's prudent to use white, grey, and black in these cases for the purpose of differentiation. Some colours have an association with different incidences like red for danger, green for go or safe, and yellow for alert. These colours have already coded information in them and users after looking at them can immediately recall what it means.

While using colours to differentiate information, it is better not to use more than five colours; else, it will be confusing to the human eye. In such cases differentiation could be done through a high colour contrast or/and different levels of colour purity.

3.6 WITH COLOUR WHERE DO WE GO?

When it comes to representing textual information through colour, there has to be a good balance between colours, typography, and the white space. We need to decide the maximum number of type family or style to be used and try to focus on using minimum colour and font styles in textual information. There is

FIGURE 3.2 Tiny geometrical shapes in colour are difficult to understand. Photo by EKATERINA BOLOVTSOVA: https://www.pexels.com/photo/geometric-shapes-of-different-sizes-7307568/.

still no "prescriptive" solution and we have to follow a holistic approach keeping in mind the different ergonomic attributes of colour usage in information design.

3.7 THE GROSS STRUCTURE OF VISUAL INFORMATION

You meet many people in life, but some of them or rather a few of them make a change in your life and you remember them forever. You might have noticed that in maximum cases the "first" impression is the most important. If you like someone in the first impression, then that's the lasting impression and you tend to like him or her forever. On the contrary, if the first interaction was not a pleasant interaction, then it is difficult to carry on the friendship for long. On a similar note, when you look at textual information in any book, brochure, pamphlet, it's the arrangement of the typographic elements, visuals, usage of colours, and prudent usage of white spaces which create an impression and you like it. The moment you like the overall "look and feel of the page", you are then interested in going deep into it and you start reading it.

You are travelling home and you hear that the train is late. You move towards a magazine stall on the platform and pick up a magazine. As you flick through the pages, you don't like it. It looks so disorganised and "ugly". The texts on the pages have no headers, no visuals, and there is no trace of organisation of text and visuals! It is a complete chaos. You keep the magazine in place and gradually walk off from the magazine stall. Why did this happen? As I said before, the people who designed the layout of the magazine never thought of the target users and have not thought ever that users do not have any compulsion to buy or read it. Through proper ergonomic design, they should attract users. So, the sequence to be followed is to first attract users to the product, and then to make them explore it, and finally to make them buy it.

3.8 ERGONOMIC ISSUES IN TYPOGRAPHY AND LAYOUT DESIGN

Humans are very choosy and are very bad at remembering things. If you present information to them at one go, their "brain" gets overloaded and they tend to forget and "loose track" of what they were reading. The human brain prefers organisation. Mr Amulya (aged 52, never been to college) goes to the nearest shop selling television sets. When he asks what the features of a particular set are, the sales person instead of explaining to him anything just hands over a brochure. When Mr Amulya opens the brochure, he is just shocked. There is information all around with no heading, no paragraph, no white space, and it looks as if someone has just typed these and expects the users to sit in the shop and spend all the time to read each and every page and then buy what he wants. Amulya, disgusted with the brochure and the content of the same of which he could not make out anything, leaves the shop. The salesperson stares at him wondering what has happened! This is a typical example where no ergonomic attributes have been used in designing the brochure. As we all know, we have a "limited RAM" also known as "short term memory". Thus, if you provide information to users without "arranging" them properly they would not be able to make any sense out of it. Thus, any textual information needs to be arranged so that the eyes can "scan" it easily and the brain can decode it fast. There are many different ways in which this arrangement of text can be done.

One of the ways to arrange is to "prioritise" information by the use of headings and sub-headings. You open your morning newspaper to find that a vaccine for giving complete protection against cancer has been invented.

That's very important news and thus becomes the main heading of the newspaper. On the same day in the same newspaper, news comes up related to some petty theft in the local shop that becomes a heading but not the main heading, it becomes heading number 2. So, as per the importance of the information, it could be arranged and thus becomes easier for users to choose what she/he wants and not force them to go through all the information. There is another way in which major information can be emphasised by using bullets and captions. It is as if the bullets and captions are saying "listen, your attention please". Normally key points in the news are emphasised by using bullets and captions in the text.

Going through textual information is like wading through a crowded street (Figure 3.3). You tend to get lost in the crowd, unless someone holds your hand, or there is certain information on the road telling you the name of the place or which direction is what landmark and so on. If nothing is there then at least there should be some people to help you out with the information that you need to navigate through the crowded road. In other words, you need some hand-holding when in a crowded road or when you are new to a place. The same thing happens when a lot of text or textual information is presented to you and your condition becomes like someone on a crowded street where you have never been. In such cases, we do have some help at hand which we can use to guide our users in steps and in the direction they would like to go. These "hand-holdings" are done through changing the "size" of the typography, using "capital" or "upper and lower case", and by using different font "families" like Times Roman, Arial, etc. You can say that these are our "rescue team" which helps the users in an ocean of typography. While trying to hand-hold users, you need to be careful not to provide them with too much guidance; else, it becomes irritating for them. For example you have gone for career counselling. One counsellor tells you to study science, another arts, and the third one commerce. You are confused what to study as you had come for some guidance regarding the course that you should pick up and instead have been further confused.

3.9 DIFFERENT WAYS OF HAND-HOLDING OF USERS IN THE OCEAN OF TYPOGRAPHY

Grouping of information in "chunks" helps in guiding the users in steps. This grouping of information can be done through different ways. You can make

FIGURE 3.3 Wading through a crowded street. Photo by Cameron Casey: https://www.pexels.com/photo/people-on-sidewalk-selective-focal-photo-1687093/.

the typography compact or make it sparse with extra spacing. In this way, it stands apart from the "crowd of normal typographic character". You can also increase the space between the lines in the text and that would also make it stand apart from the rest. In fact, you have the liberty to group by using all the mechanisms presented here together or individually. What happens in this case is as the eyes are very good in recognising any pattern, these changes are perceived as a "different pattern" in the ocean of similar patterns. These different patterns always stand apart or try to say "I am here" and draw the user's attention.

You can take the above a step forward. You can pick up the "grouped" typography and then start arranging them on the basis of the most important ones to the least important ones. In fact, you can go a step forward and offset (shift them from the normal position) and place them a little "off the track" so that they stand apart or change the size of some typography (make them bigger and smaller) to make them stand apart from the rest.

Mr Dada (aged 28) goes to the market to buy mangoes. He wants to buy different varieties of mangoes. He goes to the wholesale market, and when he looks at the mangoes, he is totally lost, no clue as to which mango is of which variety. All look the same. The salesperson in charge looking at him understood his problem. "Sir, please come to this room. I have the mangoes arranged for you so that you know exactly the variety of mangoes you are looking for". Mr Dada accompanied the salesperson to the next room. There he found the mangoes arranged in groups. On the top you see that group "those are the most expensive" ones. The next down the line as you move you come to the cheaper variety. Mr Dada could see mangoes in groups arranged in rows (Figure 3.4). In some groups, the spacing between the rows was increased compared to the rest "this indicates they are overripe". You see some mangoes standing apart on the sides? Those are the mangoes which are the ones having maximum weight in the group. There you see some mangoes kept in groups but with enough space between? They are the ones which are seedless! On the other side you see the mangoes densely grouped? They are the ones which have only seed and minimum pulp. Mr Dada was impressed; he could now decide what he wanted to buy and what not to.

Users expect certain information in certain areas of the visual field. If you have the chance of knowing this preference of users and then designing the layout of the textual and visual information becomes much more "user friendly". For example if there is a photograph, then possibly majority of users would prefer the legend for the same "below" the photograph and not "above" it. If you have to select that text on the right or on the left side of an image, then based on geographical location users would prefer text on the right side of the image (where there is no space for placing the text below the image) compared to the left side. So, if you divide the white page into sections based on users' expectation and then put text and images there, then users can read, understand very fast, and also have the option to read what they want and skip what they want to avoid.

FIGURE 3.4 Mangoes arranged as per variety and price. Photo by Shieneth Murro: https://www.pexels.com/photo/ripe-mangoes-on-the-table-4952494/.

3.10 COLOUR AND TYPOGRAPHY

You need to be very careful while using colour in typography. If you see your morning newspaper written in red, green, and blue all over then you would be a little irritated. This is where black text scores better on a white background. In books and other places of textual information you may use coloured text, but that should be to "re-confirm" or "re-emphasise" certain information. For example some specific information like formulas or a specific terminology in a book which is very important can be in a coloured typography. In a poster for a drawing competition, the last date for submission of application could be mentioned in a different colour to draw the attention of the users, because it's very important. Colour can be used (with prudence) in the beginning of a chapter, in differentiating different sections in a book or brochure, and for indicating cross references in journals. These usages of colour help the information to differentiate itself from the crowd and make it easily noticeable. That's the reason for trying to stick to one colour in typography. If too many colours are used, it's distracting. Its like kids shouting "I want chocolate!" if everyone shouts (if different colours of typography are used), then you are confusing your users as to where they should focus their attention.

When you are using colour in typography, make sure that the thickness of the letters or numerals is increased compared to black ones. Use a darker colour as lighter colour tends to merge with the white background and thus for differentiating or highlighting some textual information, it's wise to use darker colours like violet, deep green, deep blue, etc. You will also notice that as you juggle with colour in typography, a very tiny area in a particular colour appears darker than a larger area of the same colour. Thus, at times in typography while using colour, you need to increase the area to get your users to know the exact colour.

3.11 KEY POINTS

A. Humans prefer colour in information presentation.
B. We need to be prudent with the usage of colours and not overuse them.
C. When small elements are to be represented, black and white and grey scales are better compared to colour.
D. We need to strike a balance between typography, colour, and white space while designing information in print medium.
E. Alignment of typographic elements plays a key role in improving ease of reading.
F. Typographic elements need to be arranged as per importance, sequence, and frequency of usage.
G. For highlighting important textual information, use colour, font size, and spacing between the letters and numerals.
H. Play with line spacing and placing some textual materials in margins to emphasise on specific information.
I. Divide your page into sections or compartments depending on users' preference for information in expected quadrants.

3.12 PRACTICE SESSION

A. Design the label of a baby food using colour and text most prudently. The information should contain the nutritional content of the food.
B. Design the label for the name of a road where there is almost no illumination during night.
C. Design a small brochure of your design department highlighting the key strengths.
D. Design a poster for raising awareness among the common people for protecting themselves from the COVID-19 pandemic.

Visual Ergonomics in Emergency Situations and for the Challenged and Elderly

4

Overview

This chapter gives an overview of the different visual ergonomics in emergency situations which users face and which demand a different ergonomic approach. The readers are introduced to the problems faced by different types of challenged and elderly users and some ergonomic directions for the same. The different visual ergonomics principles are discussed in detail in the context of a product and its labelling.

4.1 VISUAL ERGONOMICS IN EMERGENCY SITUATIONS

There are many contexts in which visual information which you see around you might not work. This visual information was designed keeping in mind

DOI: 10.1201/9781003369516-4

normal circumstances, but when circumstances are not normal, then things or information around you might "look like something else". Mr Nitai is a middle-aged man (50 years old) who works in the local jute mill. He normally visits the hospital outpatient department for various ailments. The hospital has a good information system which helps him to navigate and reach his desired destination or the doctor's chamber or the pathology centre. One day, Mr Nitai was travelling with his friend to the nearby pub when suddenly their car met with an accident. The accident happened in front of the hospital where Mr Nitai used to visit very often. When this incident happened, Mr Nitai was absolutely bewildered. He called the ambulance and it took him to the hospital. When he entered the hospital premises, it looked unknown to him and the information system on the premises did not make any sense today. Nitai was a different man today and was in a different "state of mind" and was completely confused and puzzled. When the doctor at the casualty asked him to go to the accounts department and deposit the money for his friend's admission, Mr Nitai had to ask people where the counter was. Prior to this, Mr Nitai had been to the accounts department multiple number of times, but today he was completely lost and the information in the hospital did not help him anymore, or he was not able to understand all that was written in textual form or in the form of visuals. This is exactly an example of an emergency situation in which our senses do not work as they do during normal circumstances. Our eyes fail to pick up information and our brain fails to process them effectively. Such contexts or situations demand a different approach in the designing of visual information systems so as to ensure that they are "understood" by those under some sort of "mental stress".

4.2 VISUAL ERGONOMIC APPROACH

Presentation of information through the visual route or the eyes has its own challenges and you need to factor in many tangible and intangible elements for addressing them. Let us say that you have to lie on the ground and crawl on your hands because there is a fire in the hotel room. You have to navigate yourself out of the room by looking at the exit signs. So, what does the context demand? While designing signage inside hotels, one needs to keep in mind that fire could break out. If it does, then people might have to crawl and navigate themselves out of the rooms and come to safe places. So, the body position has to be accounted for in determining the position of some of the signage used for navigation. Thus, the visual cone of a user while standing changes while he is supine or crawling or not in an erect position and this should be factored in while placing the different information systems at different places, like exit arrows on the wall near the floor or on the floor itself.

Many times, inside public transport like buses, boats, etc., there is a lot of visual information either in the form of text, numerals, or visuals. Under normal conditions, for the user to look at them and then understand what exactly they are meant for is not a problem. But in case the vehicle is moving, then it's difficult to see, read, and decode visual information. Try reading a newspaper while inside a boat and the boat is on choppy waters! Very difficult, isn't it? Or try reading the value of the blood pressure of a patient inside the ambulance, when the ambulance is moving on a road full of potholes and is vibrating! It's difficult, isn't it? So, we need to be ready for these unforeseen circumstances, and for textual information, we might have to increase the typographic width and height both. You need to check for from where exactly you want your users to read the visual information.

You are travelling with an injured friend who is bleeding profusely. You have to reach the casualty very fast because there is no time to call the ambulance. You are trying to navigate through an unknown road. There is so much information that you are finding it difficult to locate the directions to the hospital because they seem to be hidden within the normal information on the road. Finally, you reach the hospital. In such situations, while designing information systems that are meant for emergency situations, you need to keep in mind the amount of mental load (cognitive load is the scientific term) the user is subjected to. Let's say that you are to leave your home for the railway station and are already late. The taxi is waiting for you. Suddenly you are unable to locate the keys to the main door. You start panicking and start looking for the keys all over the house but are unable to find them. Then suddenly your friend next door enters your house and locates the keys for you which happen to be kept on your dining table. This is what happens when we are stressed. Apparently easy things become difficult for us to locate. So, important visual information should be made so explicit that they are easy to locate and stand apart from the "crowd" of information.

The temperature outside is very hot. You are sitting inside your car and driving to the nearest automated teller machine. The air conditioner of your car is set to the lowest temperature. You get down from your car to withdraw some money from the automated teller machine; suddenly you are unable to see anything because of the condensation of water on the lens of your spectacles. Did you think of this? Just at that time a biker on his pushbike comes and hits you hard, because you are not unable to see anything, nor are you able to locate the automated teller machine. You take out your hanky and wipe the lens of your spectacle … This can happen even when you enter an air-conditioned room from a very hot environment and all the visual information in the room can look blurred. This is where we need to play around with increased contrast, increased typography height and width ratio, and the judicious usage of some colours with a light background and dark text which would help users identify them easily. Just imagine you get down from an air-conditioned car and because of the same reason are unable to differentiate between the male and the female washrooms because you are unable to differentiate the symbols for both specially if there is a nature's call and you have to urgently enter the toilet. The only solution to this is to make the symbols bigger and use darker and brighter colours which would make differentiation relatively easy in this context.

While driving during heavy rainfall, you are unable to see anything in front. If there is a sharp bend on a hilly road, then you will not be warned of the same as you cannot read it, and it could lead to accidents. The way out of such a situation is to use a dynamic display or a flashing display which will be able to draw your attention. Additional audio signals could be used to alert users in such circumstances.

4.3 THE CHALLENGED AND ELDERLY POPULATION

As people age, there are certain changes in the eyes which visual designers need to factor in. Cataract or yellowing of the eye lens is one such phenomenon. The users in such cases are looking through a yellow filter. Thus, if you use specific colours to convey certain information to your users, they might not see them in that colour. For example if the information is to differentiate between two sets of textual information in two different colours. This has to be factored in and if necessary "apparently useless" information should be used like using symbols/pictograms along with text to minimise the chances of missing the visual information being conveyed. Such additional or useless information at times acts as re-confirmation to your users and at times is extremely valuable.

Elderly users also have a problem at times in visualising pure blue. So, they would see anything in pure blue in shades of grey. So, if you are using pure blue as a contrasting colour along with some other colours, you need to keep this at the back of your mind. Especially if the typography is in pure blue, then some elderly users would perceive them in shades of grey and the information which was supposed to "stand apart" from the rest would not for this specific group of users. The solution to this could be to avoid using pure blue and add some impurity to the blue colour and use it thereafter.

We have learnt before (chapter on eye) about the phenomenon of light and dark adaptation of the human eyes. In case of elderly users entering a relatively darker space from a bright sunlit area, it takes a longer time to adapt to the relatively low illuminated space. Thus, the information systems which are used to communicate with the users might not be of any use to them. The way to address this issue would be to provide more time for these groups of users for adaptation. Thus, visual information for elderly users in case they are very important should not be provided at the transition point between the external and internal environment but a little inside the darker environment so that users get some more time to get adjusted. In specific exhibits where the display of visual information starts right from the entrance, some transition zone should be available where the users could adapt their eyes to the low illumination level before moving any further. Apart from the above, elderly users lose their power of looking at near and distant objects in quick succession with acute problems. This means that while they look at a distant object and then suddenly turn their eyes to a near object, and vice versa, they see them blurred. Thus, they need some time before their eyes can see information clearly. This is important in places where they have to look at display boards at a distance and after some time need to read some information very close to them. Some time needs to be given to them.

Significant portions of the world's population are red and green colour blind. This means when trying to convey visual information in pure red and pure green, these users would perceive them in shades of grey. One of the many solutions to this problem is not to use pure red and pure green but to use an impure variety of the same. This would ensure that the presented information is understood by the users in adequate contrasting colours.

4.4 VISUAL ERGONOMICS OF PRODUCT LABELLING

While buying products for our day to day use or any over the counter medicines, it's the packaging and labelling which "attracts" the users. So, the product "label" can be held responsible for the "love at first sight" of the customers. The information be it textual or exclusively in the form of images and graphics plays a very important role in ensuring this "love" between the customer and the product and there are stages in this process from the visual ergonomic perspective.

Mr Terence, who is 72 years old, needs to buy some balm as he has been suffering from a headache for the last few days. He travels to the local supermarket and enters. His first step is he "looks around" for that probable section in the supermarket where this could be available. He asks himself is it in the cosmetics section? Or is it somewhere else? This supermarket does not have a pharmacy and hence no question of enquiring about that. Finally, he asks the staff and the staff points to a place at one corner next to the cosmetics section.

Mr Terence finally manages to reach that section and finds the place which has many over the counter drugs for cough and cold, sprays for sprains, lozenges for sore throat, and pain balm. He starts "looking" for his balm and his eyes fix on one of the balms which have a unique packaging that he likes. He takes the product which is a nice, cute, and small product and starts reading it. First, the "name attracts" his eyes and next he looks for whether this could be used for a headache or not. After being satisfied with that he then looks for the price of the product and the date of manufacture and the date of expiry.

If you keenly observe the steps that Mr Terence followed, then you would notice some important visual ergonomic attributes. Though we are talking about a simple product label, it "exists" in the specific context of the supermarket. Mr Terence had to look for the product even when he was standing near it because there were other similar products and it took time for Mr Terence to "identify" the desired product from the "crowd" of other products. This was done through the labelling which had specific graphics and colour. So, colour and graphics played an important role in helping the product "stand out" from the crowd of products. As a visual designer, you can go an extra step ahead in changing the form of the product. This becomes an extra cue for your users. Let us imagine that Mr Terence was standing in front of the rack with all bottles of balm of the same size and form. If one of the bottles were of a little different shape or form it would stand apart from the crowd! So, even as visual designers, sometimes you need to use some principles of product design as well. After this, he looked for a reconfirmation of whether the product is a balm for headaches or not. Once that was confirmed, he moved to the price tag and the expiry and manufacturing date. So, users when they read information in a product, they do it in stages and have a "priority" for the information they are looking for. This need not be the case always. For example if Mr Terence buys the balm very often and knows the area in the supermarket which houses the balm, then he would walk straight to the rack and just look for his favourite brand and pick it up and go. This is because he is now an "expert user" or a person who has purchased it before. In the previous scenario, he was new to the context of buying the product. But when you apply the principles of visual ergonomics to product labelling, you need to factor in both the expert and the novice users as well.

I'll tell you another interesting story of Mr Terence relevant in this context in his narration. One day, I was about to leave for the supermarket to get my weekly groceries, fruits, and vegetables. The previous night, I and my wife had prepared a long list of the same. As I stepped out of my house and was about to get into my car, my wife shouted, Terence, please get my lipstick for me and saying this she tells me the name of the same. I replied no problem Jenney, I will get it for you. Saying this I drove to the supermarket. I completed all my purchases and finally checked out of the market. When I checked out, I suddenly remembered that I had forgotten the name of the lipstick she mentioned.

Now if I call her she would get angry as I told her that I'll remember it. I sat on the bench just outside the supermarket and closed my eyes and started thinking where does she keep the lipstick? As I closed my eyes my mind started travelling, it entered through the main door, arrived at the living room and then moved towards the refrigerator, on top of which there was a small tray where she kept the lipstick. Believe me as my mind went to the tray on the refrigerator, the name of the lipstick flashed in front of my eyes ... Without wasting time, I again entered the supermarket and purchased it. This incident of Mr Terence indicated the issue of how our brain retrieves information with reference to certain landmarks (points of reference like the refrigerator, the tray, etc.).

4.5 VISUAL ERGONOMICS OF THE BALM

It is said that the first glance at the product is the most important one and then the user starts to look into further details. With reference to the example of the balm, let us look into the visual ergonomic aspects of it (Figure 4.1).

As can be seen, it's very difficult to identify each letter that is an "i" as an "i", "t" as a "t". The spacing between the letters is so condensed that the pattern that

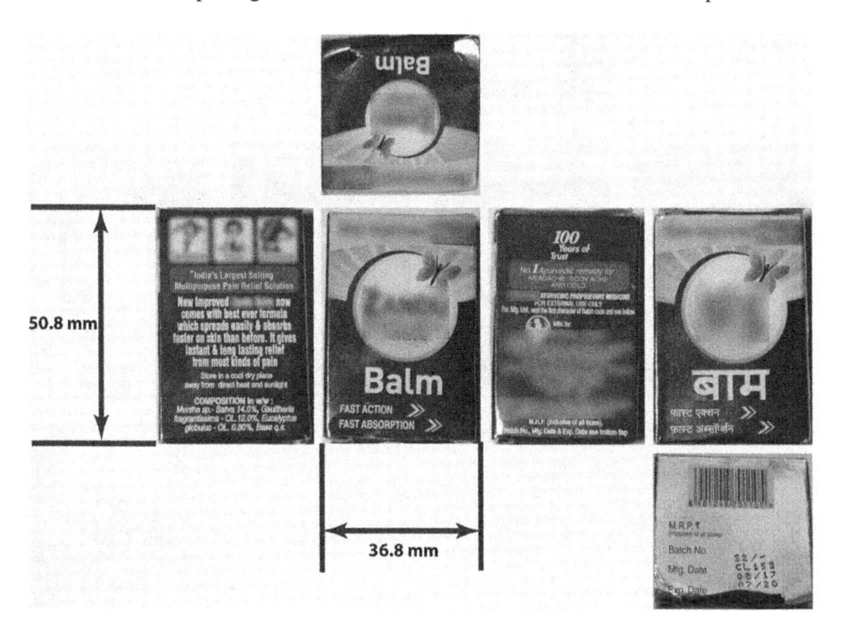

FIGURE 4.1 Visual ergonomics of the existing balm label.

the upper and lower case generates is very confusing, and adding to this is the lack of spacing between the lines. Much information was unnecessary, and prioritisation of information was lacking. Any user looking at the product labelling would first need to know the brand, and then what it's all about, followed by other details. The price, manufacturing, and expiry dates are not in the expected quadrant of the visual field and it's difficult to identify the individual "characteristics" of the letters and numerals as they are. In the date, the numeral eight can be mistaken as three and the zero can be mistaken as the letter O. The visual of the butterfly with the balm is not in accordance with the mental model of users (what the users think of balm and its relation with a butterfly). The visuals depicting the usage of the product also were not in tandem with the mental model of the users in performing the task of applying balm on the affected area and the exact way of doing that including the quantity to be used. Thus, the labelling in other words did not match the way users think and expect the product to be used. This indicated mismatch between the product and the user and demanded visual ergonomic intervention in the labelling of the product without changing the product form.

In the new labelling system of the product as shown in Figure 4.2, a colour scheme has been selected which avoids the usage of pure red, green,

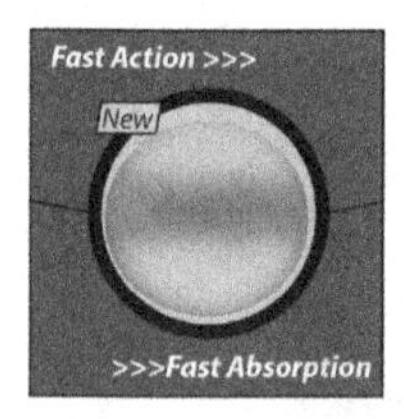

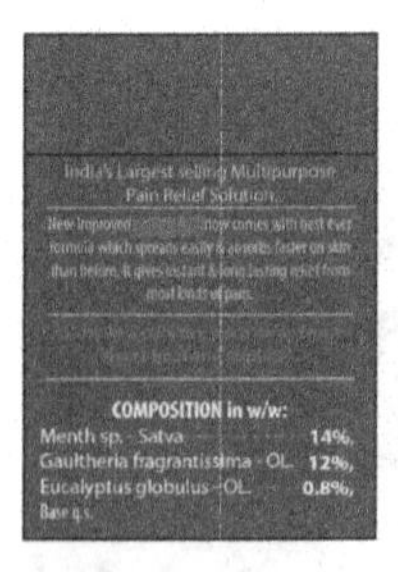

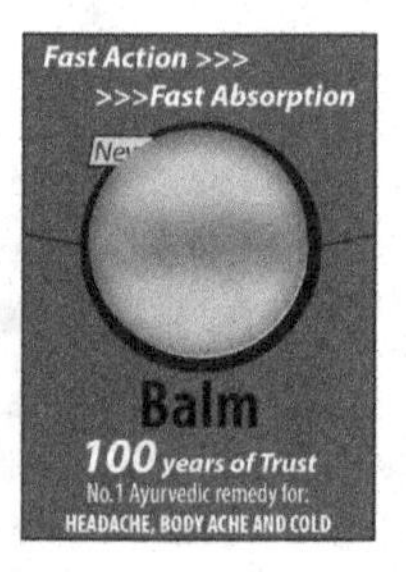

FIGURE 4.2 Visual ergonomic interventions in product labelling.

or blue colour, keeping in mind the colour-blind users (the red, green, and blue ones only, as they are the maximum in the population). The contrast of the text has been increased against a relatively darker background with text in a light colour. This ensures that the label is visible in low illumination also given the fact that the user study revealed that this product sells not only in supermarkets where illumination level is adequate but also in rural and remote rural areas in small shops where the illumination level is very low. Some of the information needed by the users (through user study) are highlighted in a manner that they "pop out" from the crowd of textual information. Some of these are "new", "fast action", "net quantity", as these are "anchor information" with reference to which the user decides whether to buy or not to buy the product. The price, batch number, manufacturing date, and expiry date are a separate group of very important and mandatory information which needs to stand apart from the rest. Thus, they are written in dark text against a grey background, with prioritising of the information to their relative importance and thus price comes first followed by others.

Figure 4.3 shows the improved label design with horizontal lines used to separate information according to their importance. The most important information is in bold and "pops up", thus guiding the user to store it in a cool

FIGURE 4.3 The improved label on the right with proper "grouping" of information and highlighting the ones that demand users' attention.

FIGURE 4.4 The front label of the product (old left and improved right).

and dry place. Some users are interested in the composition and thus it's kept at the end and it's a government regulation as well. Guide lines have been used so that users can easily get to know the percentage of each ingredient and thus a strong relationship is established between the ingredient and the percentage in the composition of the product.

Figure 4.4 shows how on the front label the main characteristic of the product is indicated; "Fast Action" and "Fast Absorption" are written in upper and lower case to generate a pattern which can be easily recognised by the users. The words "headache" and "body ache" are written in all caps because user study revealed that majority of the users buy balms on the basis of the symptoms it can relieve.

Figure 4.5 shows that horizontal lines have been used to categorise the information needed as per the importance of use. The amount "Net Qty" has been given emphasis as it's very important for users and facilitates their decision making for buying the product. The instruction "external use" is important from a safety perspective and is written in red as it's an instruction and warning to users at the same time so that they do not ingest.

Figure 4.6 shows the price, manufacturing, and expiry dates. These are again important information depending on which users buy the product. For example products nearing expiry dates won't be purchased by those users who do not use the product on a regular basis. These visual ergonomic interventions have been done on the basis of a detailed user study and after the new design, the same was validated on the target users once again to see whether they are working or not.

FIGURE 4.5 Labelling on the side of the product (old left and improved right).

4.6 VISUAL ERGONOMICS OF PESTICIDE PACKAGING

Once upon a time, there was a farmer named Mike. He was a senior citizen and stayed all alone. He had a small piece of land on which he did his farming and was happy with what he produced and sold in the local market. Unfortunately, a portion of his agricultural products was destroyed by pests and he was very frustrated with that. His friends advised him to get some pesticides from the local market and use them for getting rid of the problem. As advised, he went to the local market and tried to locate a pesticide for his affected agricultural produce. The local shop where he went was a little dimly lit and as he entered the shop, he was confused as to which pesticide to select. He started looking at the racks but was not able to decide on the right kind of product because the packets did not properly display the information about the type of pesticide and on which insects it could be used. The proper dosage was also not depicted. So, Mike was unable to take any decision and just came back home utterly frustrated. We address the problems faced by Mike in the following paragraphs.

In this exercise visual ergonomic principles have been used to analyse and redesign the pesticide packaging of a popular pesticide among the

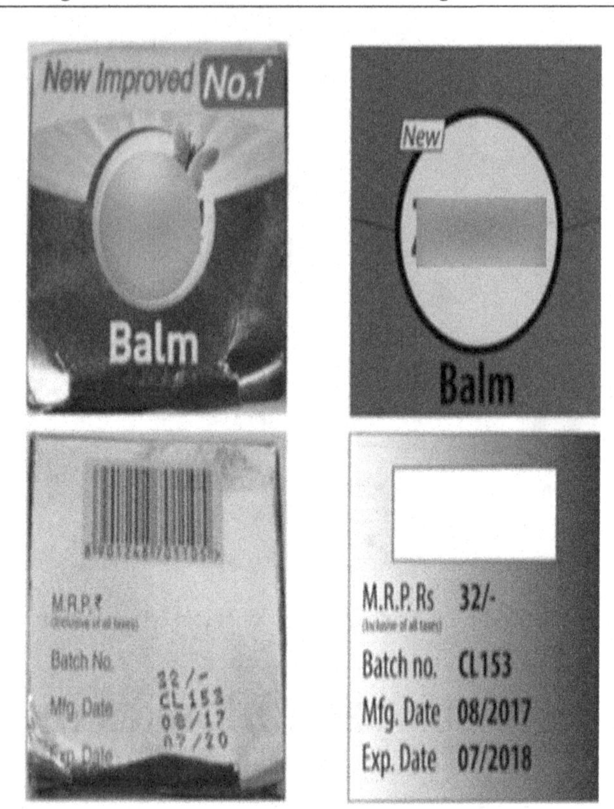

FIGURE 4.6 The price and composition label (old left and improved right).

farmers. This pesticide packaging had severe problems in terms of the overall information design and the illiterate and semi-literate farmers were finding it hard to understand much of the information which was necessary for using the product.

Figure 4.7 shows the front part of the package. There is no clue for the users what the packet contains. The visuals and the texts do not convey what is the function of the product, and what it does. The icons used at the sides are not in tandem with users' mental models (what they think and is the image conveying that). That the product is poisonous is not popping out, whereas this is the most important information to be highlighted as many farmers die because they do not take precautionary measures while handling pesticides.

Figure 4.8 shows the back of the pesticide packet. The font used makes it difficult for users to read it with ease, and the compositions and the contents are far apart and it's difficult for users to relate the composition with the content. To handle a problem like this, it's better if you look at the context of the

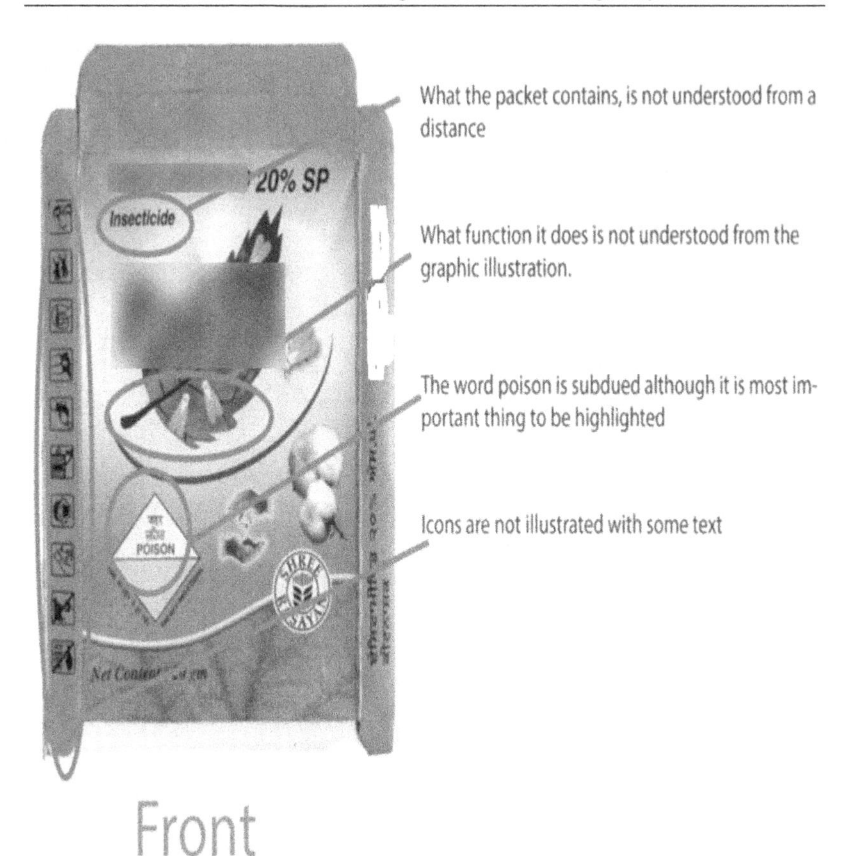

FIGURE 4.7 Front part of the existing pesticide packet.

product (in this case the pesticide packet). This product was designed on the drawing board by someone, it was then printed/manufactured, and finally it was transported and found its place in the shop, and while it was in the shop, it was again placed on the shelves. So, at every step, there are different sets of users who need to handle the product, but the end user is the one who would be using the product based upon the visual information on the packaging and this is our area of concern right now. The end user like Mike would be buying the product and then storing it and again reusing it when required. In all these tasks, Mike would be guided solely by the information on the packaging.

The user study revealed that if users like Mike are semi-literate, that is, they are able to read to a certain extent, then the information presented should be in a manner that users can understand. This is where usage of visuals and proper prioritisation of information could play an important role in guiding the

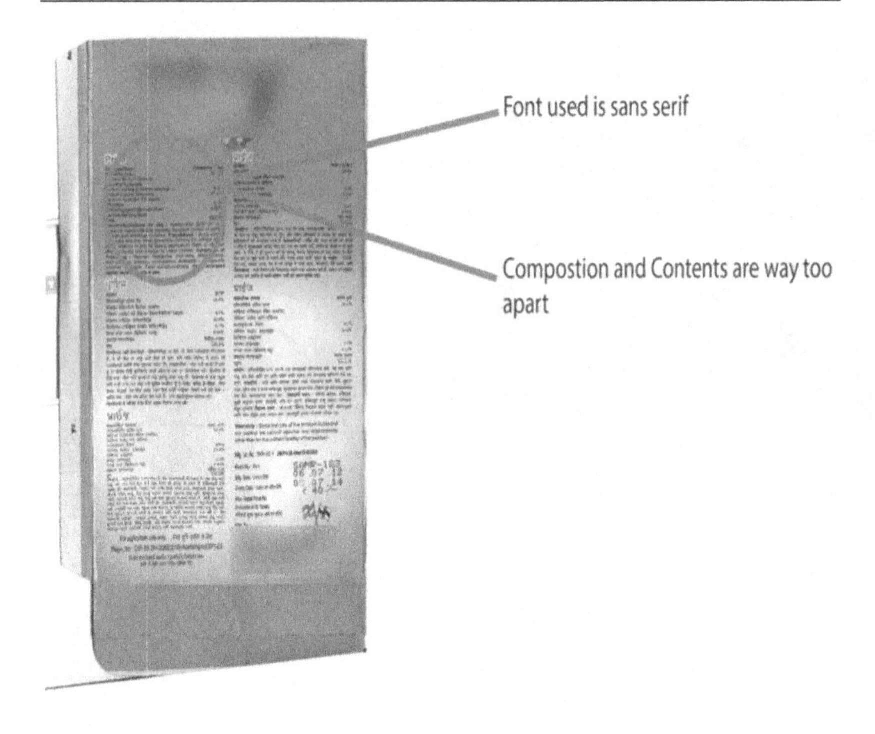

FIGURE 4.8 Back of the pesticide packet.

user in steps towards using the pesticide. The prime focus was on grouping the information in a manner which is easy to understand. Placing required information in the "expected" quadrant of the visual field of the user so that finding out the same becomes easy. Use of a "mental model" in designing icons so that decoding of coded information becomes easy. Information related to these was all obtained through a detailed user study on the target users like Mike. It was found that at the back of the packet (Figure 4.8) a lot of information was provided which was possibly not necessary and only added to the confusion of Mike. Based on these initial visual ergonomic analyses, it was decided that the hierarchy of information first needs to be decided. The user study in the context of users like Mike revealed that the sequence of information from top to bottom of the product should be like: name of the product, net content, price and manufacturing date, applicable to the type of crop, composition and strength, directions of use, and warning. The warning though listed last in the vertical sequence has to pop out.

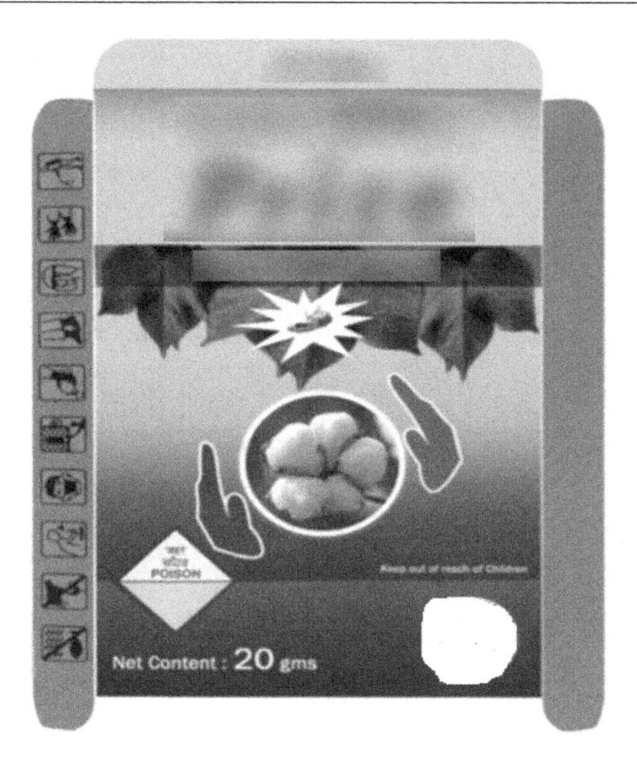

FIGURE 4.9 New concept front keeping the dimensions same as the existing one.

The users for this product would be the distributors, sellers, and buyers, so relevant information should be provided to all. The product would be used under different circumstances as it would be used under low illumination and would be kept in a dark corner of a room, on the rack, and of course in illuminated places in bigger showrooms. It will be used by a person like Mike in the open under bright sunlight and even when it's cloudy or drizzling.

The new concept of the packaging (front side) shown in Figure 4.9 was designed keeping in mind all the ergonomic issues mentioned before. The colour was chosen in a manner that it would be visible under different ambient illuminations both high and low. The information hierarchy was maintained, ensuring that users have no problem in getting the information they are looking for. Images of the crop and insects have been used to convey the purpose of the product and where exactly it is to be used. The warning information was placed with a redundant outline which is the standard norm and helps it to "stand out" from the surrounding. The quantum of information was reduced compared to the existing design making it easier

FIGURE 4.10 New concept from the back keeping the dimensions same as the existing one.

for the user to take a prudent decision of selecting the product and thereby using it in the right way Figure 4.10.

The back side of the product was designed keeping in mind the ergonomic principles. Guide lines and dark bands were used to connect the ingredients and their amount under composition. The text was aligned so as to give a neat appearance without unnecessary clutter all over. The font size of some of the text was increased to highlight its importance but at the same time to convey that it was not urgent.

Figure 4.11 shows another concept but the size of which was enhanced. The additional features in this concept were the addition of an icon for "keeping it out of children's reach", and an icon of scissors to guide the user on how and where to cut the packet. The icons depicting the product usage were given on the left side and were designed in tandem with users' mental models. Further, these icons were grouped according to their function and sequence of use to help the users perform the task much more easily.

FIGURE 4.11 New concept (with font size enhanced) front keeping the dimensions same as the existing one.

Figure 4.12 represents the same product with more spacing for the text and here the font size of the text was increased so that users could use them easily. The place to cut the product was indicated by the icon of scissors so that users are guided where and how to cut and store it for future use.

Mr Alfa, aged 45, was suffering from a fever for the past few days and decided to visit the doctor at the hospital. He took an appointment and visited the doctor who prescribed certain medications to him. The medicines were of three different types. Mr Alpha went to the hospital pharmacy to buy the medicines. As he went to the pharmacy, he found that there was a huge crowd and all the pharmacists were very busy dispensing medicines to all the patients. Mr Alfa could see that the pharmacists were very carefully reading the medicine names and then only giving them to the patients after tallying them with that written in the prescription. So, Mr Alfa thought these pharmacists are under a tremendous amount of stress because they have to dispense the medicines correctly and fast as there are patients waiting in the queue. Mr Alfa out of curiosity decided to move to the side and observe the tasks done by the pharmacists.

When a patient hands over a prescription, the pharmacist first looks at it and then reads it carefully. Then he/she goes to the medicine rack and picks up the required medicine. If the medicines come in blister packets they normally

FIGURE 4.12 New concept (with font and icon size enhanced) from back keeping the dimensions same as the existing one.

are within another box and thus that box needs to be opened to dispense the strip. If there are two medicines which have very similar names then the pharmacist has to be extra cautious so that error doesn't happen. The pharmacist he found had to waste a lot of time in case the box was turned the other way round and looked for the label. In fact, he noticed that there were some boxes containing a few blister strips where the medicine name on some of the faces was not symmetrical but upside down and in case the box is turned, it's difficult to read the medicine name. Another "grey area" in this dispensing process was that the medicine of the same type but different dosages had exactly the same packaging and the pharmacist had to be extremely careful before dispensing the medicine. Many times, the elderly pharmacist was making a mistake of reading 1.0 mg as 10 mg and was scolded by the chief pharmacist. Lastly, some tablets were sold not as a complete blister pack but in smaller units of one to two. Patients were complaining that they are unable to see the expiry date after they go home if the tablets are cut like this. Mr Alfa took his medicines but decided that he would use visual ergonomic principles to solve some of these problems to make the life of the pharmacist a little easier. A careful analysis of the above story indicates that the labelling in medicinal packaging has visual ergonomic issues. There were three problems in the story above.

Medicines with similar names demand more attention as they are more likely to cause an error. For example two medicines named "Khichuri" and "Khichuni" are very similar. As per our learning before, an upper lower case generates a pattern which is easy to ready and decode. But here the pattern generated is similar. Under stress, mistakes can happen. So, you can capitalise the "words" that are different. So now if it's written as khiCHURI and khiCHUNI then these all caps shout aloud and warn "take care I am CHURI and not CHUNI or vice versa".

The next problem was the same medicines but different dosages. Here you can use colour on the dosage and that colour helps in differentiating between different dosages. For the problem with elderly users, don't use 1.0 mg, instead write it without the zero as 1 mg, so no more confusion about mistaking it as 10mg!

To save searching time for medicines, the names of the medicines should be written in the same order on all sides so that they are detected from all angles. Lastly, the problem that users were facing when one tablet was cut from the main strip can be sorted out by having the expiry date of each tablet behind. It's expensive but solves the problem (Figure 4.13).

4.7 KEY POINTS

A. Keep in mind body posture, environmental hazards while designing visual information.

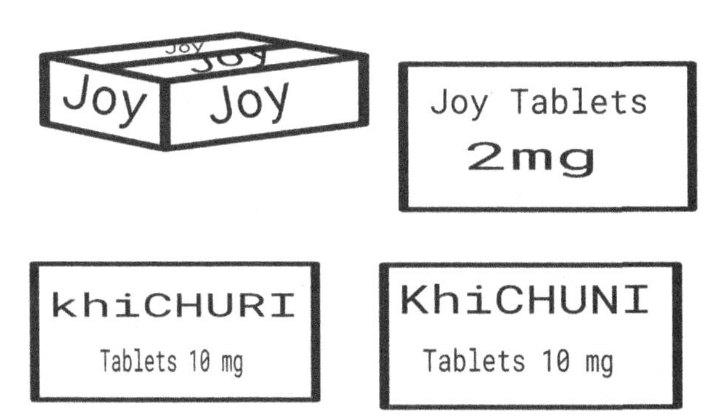

FIGURE 4.13 Visual ergonomic problems and intervention in medicinal packaging.

B. Colour-blind users are to be factored in and hence it's prudent not to use pure red, green, and blue colours.
C. People in emergency situations get tense; information should be simple and easy to understand for them.
D. Important visual information should always stand apart from the rest so that users can identify and use them with relative ease.
E. For elderly users, keep in mind that they are looking through a yellow filter, cannot always see pure blue colour, and have a problem when transitioning from bright to relatively dark areas.
F. Product form and visual ergonomics of the product label should be seen as one entity.
G. The important information should stand apart from the crowd of information.
H. Use all caps and upper and lower case judiciously, as an upper lower case is easy to read and all caps indicate very important message requiring immediate attention.
I. Be wise in using textual materials and do not overload your users with too much information.
J. Factor in the context of the use of the product while designing the product label.

4.8 PRACTICE SESSION

A. Design the male and female symbols for elderly population so that they do not confuse between the two at day and at night.
B. Design a menu card for a restaurant keeping in mind the elderly and red/green colour-blind population.
C. Design the symbol for the boarding gate at an airport for the elderly population.
A. Design the label for a pain-relieving gel to be used by illiterate users who cannot read or write. You can only use images for the same.
B. Incorporate visual ergonomic features on the tablet strip for hypertension for elderly population who forget to take their medicines and often forget whether they have taken their medicine for the day and take another one.

Visual Ergonomics of Information Collection and Dissemination to Users

5

Overview

This chapter introduces the readers to the visual ergonomic attributes to be used in designing of "forms" for collecting specific information from the target users. The challenges in filling up the form and that of retrieval of information from the same are discussed. An example is given to depict the application of visual ergonomic principles. Readers are given a glimpse of how a user navigates in a large space and what are her/his expectations in terms of information and reduction of uncertainty while navigating in spaces. How do we give reconfirmation to the users and do some hand-holding when they need the most is discussed here. The chapter ends with visual ergonomic application in simple map design for the users and how maps could be made much more user friendly.

5.1 INFORMATION COLLECTION FROM TARGET USERS AND THE CHALLENGES

Mr and Mrs Longa aged 75 and 70 years, respectively, wish to travel to the mountains for their summer holidays. They plan to take the train to the hill

DOI: 10.1201/9781003369516-5

FIGURE 5.1 Elderly couple in a booking office.

station named "Dichkao". They have to follow a procedure which is lengthy for this elderly couple because a form needs to be filled up in detail. The couple decided to go to the booking office together so that it would be easier to fill up the form together than if either of them had been there (Figure 5.1). The couple reached the booking office to find a small queue there. They picked up a form and decided to fill it up. For filling out the form, they were looking for some horizontal surface and they could find none. So, Mr Longa decided to purchase a magazine from a stall outside the office (it was just for use as a surface to fill up the form. Wastage of money!). After taking the form, the couple was confused as there were so much of details to be filled out. They were expecting that they just needed to write in their names and the destination. When they glanced through the form, they found so many columns and blank spaces that it was very difficult for them to fill the form without any help (Figure 5.2). The place inside the office the couple felt was a little less illuminated and it was difficult for them to read the form.

This is the problem today when it comes to filling up of forms by the general public for different purposes. It could be for railway reservation, income tax return, opening of a bank account, just to name a few. The example below depicts such a scenario faced by Mr and Mrs Longa and indicates how visual ergonomic principles could be applied to make the form much more "usable" by the common user. When it comes to a railway reservation form as we have seen there are two sets of users. One set is the passengers and the second set the booking clerk who needs to retrieve the information and re-enter them into the system. So, our aim here is to ensure that information entry and retrieval both become easy for all.

Figure 5.3 shows the first part of the form which needs to be filled in only by doctors and senior citizens. We agree that this is important but considering the number of people travelling, it was felt that this was possibly not the right quadrant to place this information. Our user study revealed that doctors and senior citizens were few in number compared to the other people travelling. When a person is taking the form from the counter, she/he already knows what form it is and thus writing the name of the form in upper lower case is better and using all caps is unnecessary as it only adds to the noise in the textual information already displayed. Whether a person wants to be upgraded need not be in bold as it doesn't demand so much an emphasis right now. It's important information, but there are more important pieces of information compared to that for example boarding and destination station names. The details related to the train should be given the highest priority. If you look at the context of use of filling up the form, a majority of the users (user study revealed) read the board displaying the different train details and from there they try to memorise the train name and number and then type that. Now if they are bombarded with unnecessary information before that (as such we are bad in our RAM capacity or short-term memory), they tend to forget and write the wrong train name/number or both. So, the mental model in task performance of the user's needs to be respected and the sequence of information in the form should respect that as well.

Figure 5.4 shows some ergonomic issues in this part of the form which are good. The name of the applicant (passenger), address, telephone number, and signature are grouped together with onward/return journey to ensure that the passenger details are not lost when the booking clerk tries to retrieve

RESERVATION / CANCELLATION REQUISITION FORM

If you are a Medical Practitioner Please tick (✓) in Box
(You could be of help in an emergency) Dr. ☐

If you want Sr. Citizen concession, please write "YES" / "NO" in box
(If Yes, please carry a proof of age during the journey to avoid inconvenience of penal charging under extant Railway Rules.) ☐

Do you want to be upgraded without any extra charge ? Write YES/NO in the box. ☐
(If this option is not exercised, full fare paying passengers may be upgraded automatically.)

Train No. & Name ________ Date of Journey ________
Class ________ No. of Berths / Seats ________
Station From ________ To ________
Boarding at ________ Reservation upto ________

Sr. No.	Name in Block Letters (not more than 15 letters)	Gender M/F	Age	Concession/ Travel Authority No.	Choice, if any
1.					Lower/ Upper Berth
2.					
3.					Veg./ Non-Veg. (for Rajdhani/ Shatabdi Express only)
4.					
5.					
6.					

CHILDREN BELOW 5 YEARS (FOR WHOM TICKET IS NOT TO BE ISSUED)

Sr. No	Name in Block Letters	Gender M/F	Age
1.			
2.			

ONWARD / RETURN JOURNEY DETAILS

Train No. & Name ________ Date of Journey ________
Class ________ Station From ________ To ________
Name of Applicant ________
Full Address ________ ________ ***Signature of the Applicant / Representative***
Telephone No. If any ________ Date ________ Time ________

FOR OFFICIAL USE ONLY

Sr. No. of Requisition ________ PNR No. ________
Berth/Seat No. ________ Amount collected ________

________ *Signature of Reservation Clerk*

NOTE
1. Maximum permissible passengers is 6 per requisition.
2. One person can give one requisition form at a time.
3. Please check your ticket & balance amount before leaving the window.
4. Forms not properly filled in or illegible shall not be entertained.
5. Choice is subject to availability.
6. Passengers booked on single ticket may or may not get compact accommodation in the upgraded class.

FIGURE 5.2 The original form to be filled up.

COM. 744/F.
Rev 06 (RB)

RESERVATION / CANCELLATION REQUISITION FORM

If you are a Medical Practitioner Please tick (✓) in Box
(You could be of help in an emergency) Dr.

If you want Sr. Citizen concession, please write "YES" / "NO" in box
(If Yes, please carry a proof of age during the journey to avoid inconvenience of penal charging under extant Railway Rules.)

Do you want to be upgraded without any extra charge ? Write YES/NO in the box.
(If this option is not exercised, full fare paying passengers may be upgraded automatically.)

Train No. & Name ______ Date of Journey ______
Class ______ No. of Berths / Seats ______
Station From ______ To ______
Boarding at ______ Reservation upto ______

Sr. No.	Name in Block Letters (not more than 15 letters)	Gender M/F	Age	Concession/ Travel Authority No.	Choice, if any

FIGURE 5.3 Portion of the reservation form showing information to be filled by selected few.

information from the form later on. The lower part of the form labelled "official use only" should be placed in a different quadrant; else, passengers might make the mistake of filling it out. It lacks segregation from the rest of the elements.

The lowest part of the form (Figure 5.5) has a lot of notes which take up space and make the area look very crowded. This might lead to a wastage of time by the passenger who has nothing to do with them. Thus, segregation of the notes is warranted in this case, as this is occupying an important quadrant of the form where other important information could be included. This part of the form has a few ergonomic issues. The words to, from, reserved up to, boarding at are very confusing as they mean almost the same. Users make a mistake in the onward and return journey as the information is all mixed up. In fact, foreign tourists have no guidance on how to write the class of travel in terms of its code as it's not mentioned anywhere.

The same form is used for reservation and cancellation which is not properly communicated and users are often confused. This part needs to be highlighted in detail. The choice for berth and meals is given in such a way that makes customisation for every passenger not possible. Whatever choice you select is applicable for all the passengers.

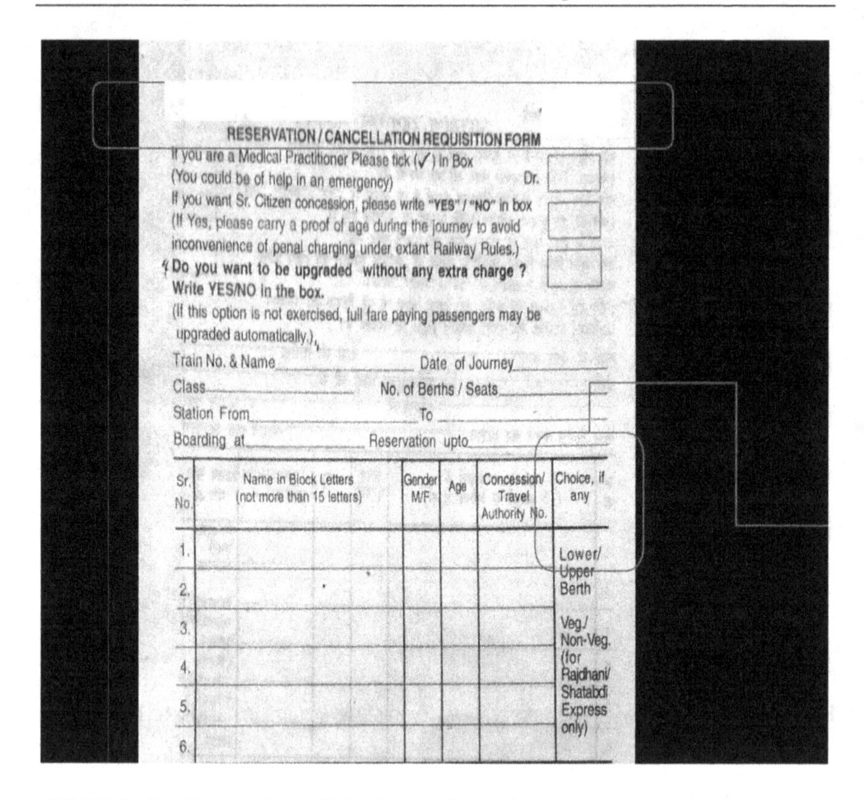

RESERVATION / CANCELLATION REQUISITION FORM

If you are a Medical Practitioner Please tick (✓) in Box
(You could be of help in an emergency) Dr.

If you want Sr. Citizen concession, please write "YES" / "NO" in box
(If Yes, please carry a proof of age during the journey to avoid inconvenience of penal charging under extant Railway Rules.)

Do you want to be upgraded without any extra charge ? Write YES/NO in the box.
(If this option is not exercised, full fare paying passengers may be upgraded automatically.)

Train No. & Name ______ Date of Journey ______
Class ______ No. of Berths / Seats ______
Station From ______ To ______
Boarding at ______ Reservation upto ______

Sr. No.	Name in Block Letters (not more than 15 letters)	Gender M/F	Age	Concession/ Travel Authority No.	Choice, if any
1.					Lower/ Upper Berth
2.					
3.					Veg./ Non-Veg. (for Rajdhani/ Shatabdi Express only)
4.					
5.					
6.					

FIGURE 5.4 The portion of the form where the journey route and passenger details are given.

When the filled in form is submitted at the counter for booking the ticket, the booking clerk enters the details (from the form) on his personal computer in the sequence such as train number, date, class, number of passengers travelling, travelling station from and to, mode of payment, type of concession, prints the ticket (the system automatically assigns seat numbers), hands it over to the passenger, and collects the money. Based on these visual ergonomic issues, some concepts were suggested.

Figure 5.6 shows that there is a clear mention of the type of reservation or cancellation which is the first information the user needs to know and with a tick mark box which is easier to select rather than striking out. This is followed by the train details and the passenger details where options for berths can be customised for every passenger. Meal preference has been removed as it's applicable on only two trains. We recommend a separate form for those two trains. Serial number and name columns have been merged, thus creating

6.					only)

CHILDREN BELOW 5 YEARS (FOR WHOM TICKET IS NOT TO BE ISSUED)

Sr.No.	Name in Block Letters	Gender M/F	Age
1.			
2.			

ONWARD / RETURN JOURNEY DETAILS

Train No. & Name ____ Date of Journey ____

Class ____ Station From ____ To ____

Name of Applicant ____

Full Address ____

____

Signature of the Applicant / Representative

Telephone No. If any ____ Date ____ Time ____

FOR OFFICIAL USE ONLY

Sr. No. of Requisition ____ PNR No. ____

Berth/Seat No. ____ Amount collected ____

Signature of Reservation Clerk

NOTE

1. Maximum permissible passengers is 6 per requisition.
2. One person can give one requisition form at a time.
3. Please check your ticket & balance amount before leaving the window.
4. Forms not properly filled in or illegible shall not be entertained.
5. Choice is subject to availability.
6. Passengers booked on single ticket may or may not get compact accommodation in the upgraded class.

FIGURE 5.5 The portion of the form with notes.

more space for writing one's name. Name in block letters has the word written in block letters as NAME IN BLOCK LETTERS acting as a reconfirmation to the users.

Figure 5.7 shows that the onward and return journey has been made clearer so that users have no doubts. Applicant details have been reconfirmed by two icons of cross for doctors and the place icon for address. The section on official use has been grouped separately with a border to separate it from the rest of the text and to ensure the passenger does not fill it up by mistake. A box for signature guides the user to sign within the stipulated space and not go beyond that (Figure 5.8).

The lowest part of the form has been used for explaining the concepts of station to and from and boarding at. This reduces the comprehension time of

☐ General RESERVATION ☐ TATKAL ☐ CANCELLATION

Train No. & Name ____________

Date of Journey ________ Class ____ No. of Berths/ Seats ____

Station From ____________ To ____________

Boarding at ________ Reservation upto ________

	Name in BLOCK LETTERS (not more than 15 letters)	M/ F	Age	Concession/ Travel Authority No.	Berth Preference
1.					
2.					
3.					
4.					
5.					
6.					

CHILDREN BELOW 5 YEARS (FOR WHOM TICKET IS NOT TO BE ISSUED)

	Name in BLOCK LETTERS	M/ F	Age
1.			
2.			

FIGURE 5.6 Improved design of the upper part of the form.

the users and makes it easier to understand as well. "From" means the station from where the train leaves and you want to board it from there. "To" means where the train is going. "Boarding" station means that if the train starts from a particular station but if you are boarding from a different station, then the "boarding" station needs to be filled up.

ONWARD/ RETURN JOURNEY DETAILS (IF APPLICABLE)

Train No. & Name ____________________

Date of Journey ________ Class ____ No. of Berths/ Seats ____

Station From ____________ To ____________

APPLICANT DETAILS

Name of Applicant ____________________ if doctor ☐

Full Address ____________________

____

Telephone No ____________ Signature (Applicant/Representative)

FOR OFFICIAL USE ONLY

Sr. no. of Requisition ________ PNR No. ________

Berth / Seat No. ________ Amount collected ________

Signature of Reservation Clerk

NOTES

1. Maximum permissible passengers is 6 per requisition.
2. One person can give one requisition form at a time.
3. Please check your ticket & balance amount before leaving the window.
4. Forms not properly filled in or illegible shall not be entertained.
5. Choice is subject to availability.
6. Passengers booked on single ticket may or may not get compact accomodation in upgraded class.

Reserve and pay

A FROM — B Boarding station (optional) — C TO

FIGURE 5.7 Improved designs, the lower part of the form.

Signature of Reservation Clerk

NOTES

1. Maximum permissible passengers is 6 per requisition.
2. One person can give one requisition form at a time.
3. Please check your ticket & balance amount before leaving the window.
4. Forms not properly filled in or illegible shall not be entertained.
5. Choice is subject to availability.
6. Passengers booked on single ticket may or may not get compact accomodation in upgraded class.

Reserve and pay

A FROM — B Boarding station (optional) — C TO

(a)

☐ General RESERVATION ☐ TATKAL ☐ CANCELLATION

Train No. & Name ______

Date of Journey ______ Class ______ No. of Berths/ Seats ______

Station From ______ To ______

Boarding at ______ Reservation upto ______

Name in BLOCK LETTERS (not more than 15 letters)	M/F	Age	Concession/ Travel Authority No.	Berth Preference
1.				
2.				
3.				
4.				
5.				
6.				

CHILDREN BELOW 5 YEARS (FOR WHOM TICKET IS NOT TO BE ISSUED)

Name in BLOCK LETTERS	M/F	Age
1.		
2.		

ONWARD/ RETURN JOURNEY DETAILS (IF APPLICABLE)

Train No. & Name ______

Date of Journey ______ Class ______ No. of Berths/ Seats ______

Station From ______ To ______

APPLICANT DETAILS

Name of Applicant ______ if doctor ☐

Full Address ______

Telephone No. ______ Signature (Applicant/Representative)

FOR OFFICIAL USE ONLY

Sr. no. of Requisition ______ PNR No. ______

Berth / Seat No. ______ Amount collected ______

Signature of Reservation Clerk

NOTES

1. Maximum permissible passengers is 6 per requisition.
2. One person can give one requisition form at a time.
3. Please check your ticket & balance amount before leaving the window.
4. Forms not properly filled in or illegible shall not be entertained.
5. Choice is subject to availability.
6. Passengers booked on single ticket may or may not get compact accomodation in upgraded class.

Reserve and pay

A FROM — B Boarding station (optional) — C TO

(b)

FIGURE 5.8 (a) Explanation of station to and from and boarding at through icons. (b) The complete form.

5.2 LESSONS LEARNT

An apparently simple product like a railway reservation form needs to be analysed from a macro and a micro perspective. The context of product usage and the target user are the two important elements we need to consider. Always try to simplify the information and try to present the information as per the sequence of actions in real life. When you plan your travel and need to book your ticket by train, you need a series of information in a sequence. When the information that you need and the design of the product are in tandem with what you need, then the task becomes easy for users. The couple planned for the hill station. After this, it's normal they would like to visualise how they would travel, how the booking is to be done, and the probable information to be written in the form. This visualisation needs to be mapped on the form.

In this example, we have seen the first task was to identify the form that it's for what specific purpose because there are many actions related to reservation and one of them is cancellation. Our user study revealed that the forms are not always given from the counter, but the user has to pick them up from the rack, and this is where identification of the correct form is important. The same form was used by two sets of users, the traveller and the staffs at the railway booking office. So, your information for both should be segregated; else, it would not only be confusing but difficult for both to find out which information is for whom. Users are humans, and humans make mistakes. Visual ergonomics ensure that these chances of making mistakes are reduced through ergonomic design interventions like providing check boxes for selection (instead of striking out unnecessary elements), making the concepts clear with graphics (doctor and cross, boarding station, etc.), and providing a box for signing so that they do not sign elsewhere.

5.3 USERS IN SPACES AND THE UNCERTAINTIES

When you move in any space, you move with a certain objective in mind. Either you just loiter in the space or see what's there all around. This happens when you enter a garden for the first time. You have time at hand and you just move around. In this case, a few pieces of information are important for you like amenities and facilities (washroom, drinking water, etc.) and other exits from the garden. The next information that you need are the names of different trees and plants and in many cases their medicinal value or what are their health benefits. If these pieces of information are conveyed to you, then your entry and trip around the garden becomes very enjoyable. Now let's say you enter a museum and are confused as to from where you start and where should you end, what are the different sections that you should visit, what each artefact is, what is its history, etc. So, in a place like this you need much more information and as a user; if you are left on your own, then you are confused and lost in the space and you are not able to "connect" yourself with the space. That means while a user moves in any space be it big or small, we need to help the user connect with the space through a proper information system. This information could be in the form of giving users only directions "move this way" or information related to amenities and facilities in space like "Toilet for male" or could be a brief description of an artefact in a museum, a plant in a garden, a particular historical monument, etc. These pieces of information in space help in removing uncertainties among the users and we don't leave them to wonder what it is all about, where to go, how to go. Thus, these pieces of information in space are just like the users' "friend" ready to help them with what they need and guide them in that direction.

Maps play a very important role in navigation. Remember that users cannot handle too much and complicated information. So, maps have to be simple and depict the real world. If necessary, maps have to be supplemented with pictures from the real world. While using textual information on maps, make sure that the text and the area of the map look like they are related. For example names of the roads should be on the road itself. The name of the mountain (Figure 5.9) could be along the slope of the mountain rather than the base. Similarly, when two rivers (Figure 5.10) run parallel and then cross, label them along the length of the river twice, once before they cross and another after they cross.

FIGURE 5.9 Labelling of the mountain name.

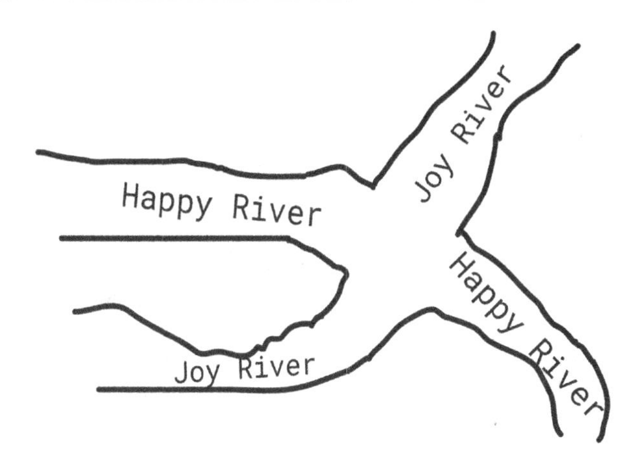

FIGURE 5.10 Labelling of the river name.

5.4 THE STORY OF MR AFU WHO WANTED TO CORRECT HIS NAME IN THE TAXATION CERTIFICATE

Mr Afu is an elderly person of 65 years of age and runs a small shop in a semi-urban area. As per the rule of the land, he pays a tax to the government based on his sales every year. It was not a problem to date, but this year after he paid his tax, he noticed that his name and date of birth have been entered incorrectly. He was told that this mistake needs to be corrected and for that he needs to visit the tax office in town and meet the taxation officer in charge of his area who works there. If this was not done, then Mr Afu would be in trouble and might have to pay a heavy fine for paying taxes with an incorrect name and date of birth. Without wasting any time, Mr Afu decided to visit the tax office in town and sort out the problem. He boarded the bus for the town and after asking the bus driver got to know that to go to the tax office, he needs to get down at the bus stop called "Agarpara". From there it's around ten to fifteen minutes' walk to the tax office. He sat on the bus and after around two hours reached the Agarpara bus stop. He got down at the bus stop and found that the streets deserted. Mr Afu was in a fix now. He wondered, "which direction should I go? Should I go right or left or straight?" There was no information on the road, nor were there any people to help him out. It was bright and sunny and the temperature was very high. Mr Afu started sweating. He thought, "let me move to the right" because "right" is always the "right" path! He started moving to the right and after walking for around 20 minutes he came across a person standing on the road with a pet. He asked the person about the tax office. The person told Mr Afu that he has come in the wrong direction and should again go back to the bus stop from where he came and from there he should move towards the left. Mr Afu again walked down the road to the bus stop and from there started walking towards the left. After some time, he finally located a big building and started looking at it very confused. There was nothing written on the building. Fortunately, he could see a police vehicle moving and the vehicle on seeing him standing stopped. "Hello, do you need any help?" the policemen from the vehicle asked. "Yes, I am looking for the tax office, could you please help me?" Mr Afu replied. "That building" and saying this the policemen pointed towards the building that Mr Afu was looking at. "Thank you", Mr Afu said and he crossed the road and moved towards the building. The entrance to the building was from the other side and it was there the name of the building "Tax Office" was written in bold. Mr Afu entered the office premises and found the area very dimly lit. It was very difficult for him to see around as he was supposed to go to the "correction section" for meeting the person concerned. After about half an hour of standing at the place, he was able to see around the dark area and could locate the staircase for the building, as the only elevator in the building was non-functional. He started climbing the staircase and reached the first floor of the building. He stopped at the first floor as he was exhausted and needed some rest. He started looking around and had no clue what offices this floor had as there was no information for him. Possibly people don't climb staircases except when the elevators don't function. All of a sudden, he could see a person come out of one of the offices with some files in his hand. Mr Afu rushed to the person and asked him, "excuse me, could you please let me know in which floor the correction section is?" "It's on the fourth floor", the person replied. Mr Afu was shocked that he had to climb three more stairs. After some rest, Mr Afu started climbing and after reaching every floor, he took a rest and looked around and finally when he came to one floor he had forgotten how many floors he had climbed. There was no information related to the floor number. So, Mr Afu again stopped on one floor and asked a person. He was shocked to know that he was on the fifth floor of the building which comprised seven floors. Mr Afu again started climbing down and this time it was easy with relatively little exhaustion and he counted every floor he climbed down and finally came to the fourth floor, his desired floor. As he stood on the fourth floor, again there was no information as to where the "correction section" was. He started moving here and there and found that the doors of all the offices were open. The nameplates of the office or the person was written on the door. As the doors were open, he could not see the nameplates from a distance but had to go near every room and had to read the nameplate. Finally, he could locate the correction section and entered the room. His job was done and Mr

Afu was very happy. He came out of the room but was again confused and was unable to find the route to the staircase again. Suddenly someone told him, "hello sir, the elevator is working. Please take that." Now where is the elevator? Mr Afu had no clue. Again, after roaming here and there, he could locate the elevator and finally came out of the building tired and frustrated.

This example of Mr Afu depicts clearly how users get lost in space if there are no proper information systems. When a space is designed, one should think of how users would communicate in space and navigate themselves without the need for asking anyone else. After getting off the bus, there should have been some information about the tax office and its direction, because this is a government office and people would need to go there often. Then, as he was moving down the road, he could see the tax office from the side and there were no labels or information on the building. Ideally, users can approach a building from any direction, and hence, buildings need to be labelled from all probable directions. It might be on the building or might be some "directional signage" on the road pointing towards the building. As Mr Afu entered the building from bright sunlight, he was unable to see anything inside as it was dimly light. We have learnt about light adaptation before. Mr Afu's eyes were light-adapted and his pupils were small. Thus, the area should have been brightly lit because it's obvious that people would enter from the bright light area outside to this place. The next problem that the elevator was not operational should have been factored in. Things do go wrong, especially mechanical ones. There should have been a glimpse of the building indicating which floor has which offices. This gives users an overall idea about where they can go and how. When Mr Afu climbed the staircase to reach a floor, there should have been some information on which floor he is. This is important information as no one counts the number of floors he is climbing. If the person was using the elevator, it would have been easy, but in that case, when the person comes out of the elevator, he needs a reconfirmation that he is on that particular floor and thus the floor number should be prominently displayed on the wall. When Mr Afu was on the desired floor, there should have been information about the different offices on the floor and their location. This would have made his life easy. Mr Afu had to go in front of every office to read the nameplate. The doors of the offices were open. This is normal and the context should be factored in. Thus, information regarding the office should be placed not on the door but jutting out from the door frame so that irrespective of whether the door is open or close, it could be seen and from a distance also. The last problem faced by Mr Afu was he was unable to locate the staircase and the elevator while coming out of the office. This happens as we have bad in short-term memory as we have seen before. Thus, every information system that guides the user in a space should be bidirectional, it should guide you inside and it should guide you to come out of the space from where you started.

Always remember that as you move in space, you are keen to know how far your destination is and how long it would take to reach that point. You also need to know how far you are from where you started and where are you now. Next comes information related to the washroom, drinking water, emergency exits, etc. These all should be depicted in the space in a manner as if they are saying "I am here". If users ask others for helping them with navigation, then the information systems in space are not working. But always remember, we are humans and as humans, our reliability is always high. So, users would always ask users for double reconfirmation while they are navigating in an unknown space. But your information system from a visual ergonomic perspective should reduce their dependence on humans. As I move, I need hand-holding at every step and information in the vicinity should take care of that.

5.5 THE STORY OF RAMU WHO WANTED TO ATTEND A WORKSHOP ON DESIGN

This is the story of Ramu who wanted to attend "Design Workshop" being held at an institute. The institute was spread over a huge 250 acres of land and navigating to the exact venue was challenging for every newcomer to the place. So, the students of the institute designed a map using visual ergonomic principles so that whoever comes to the campus can easily move to the venue for the workshop without any help.

Figure 5.11 gives the Ramu a mental model of the campus with a few landmarks and the crossroads are indicated with circle. The dot helps in identifying the venue with respect to other structures in space. The foot print icons convey how and from where to enter the venue.

Figure 5.12 shows the exact location of the information boards for the users coming to the campus for the event. The boards are placed in the cone of vision of the users so that whichever direction they come from, the information boards would be visible to them. So, with the map in hand and the information boards on the road, navigation to the venue would be easy.

Figure 5.13 shows the building and the two entrances through which users can enter the venue. To make this clear, actual photographs are also placed at the side with the information boards depicted in red and the position of the boards is indicated by a dot on the map.

Figure 5.14 indicates that as Ramu navigates and moves closer to the venue, this is what he would see along the path, which is indicated by the

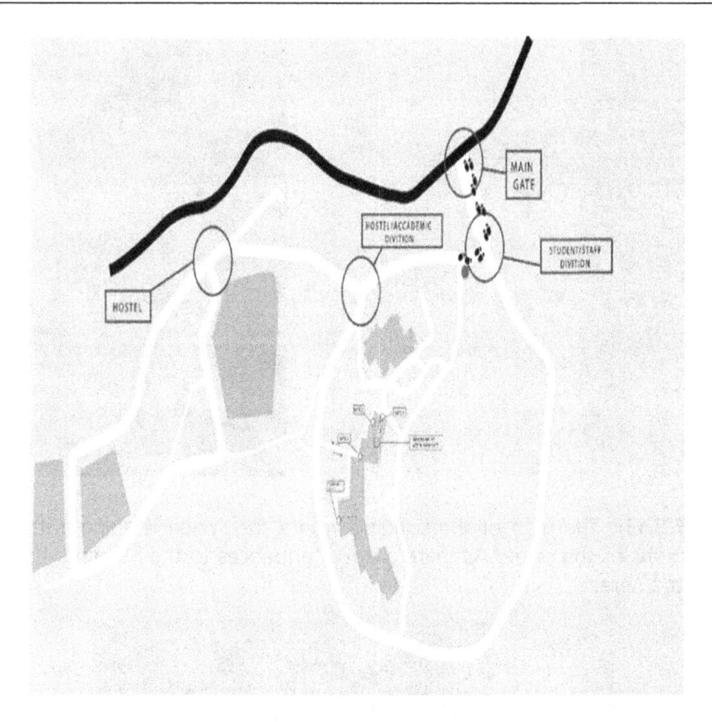

FIGURE 5.11 A map showing the overall campus and the buildings highlighting the venue with a dot. The footsteps indicate the entrance to the venue.

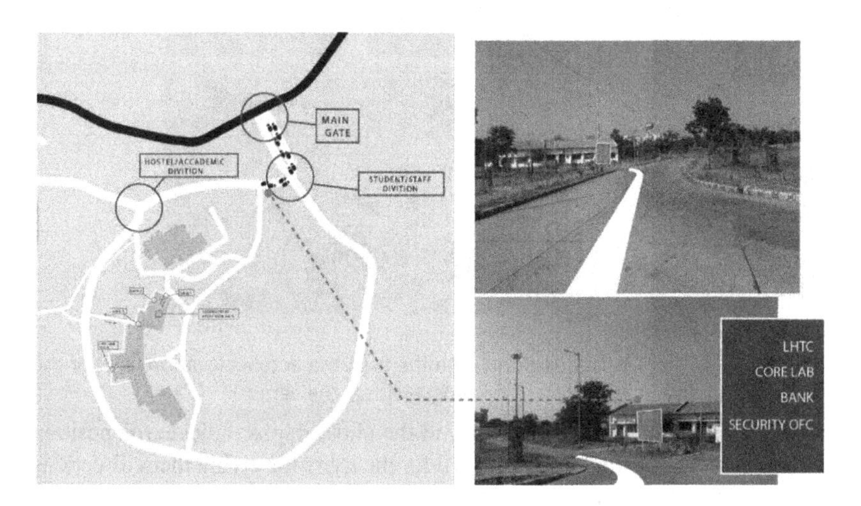

FIGURE 5.12 Shows in real space the exact position of the information boards which guides the users to the venue. The board's position is marked in dark.

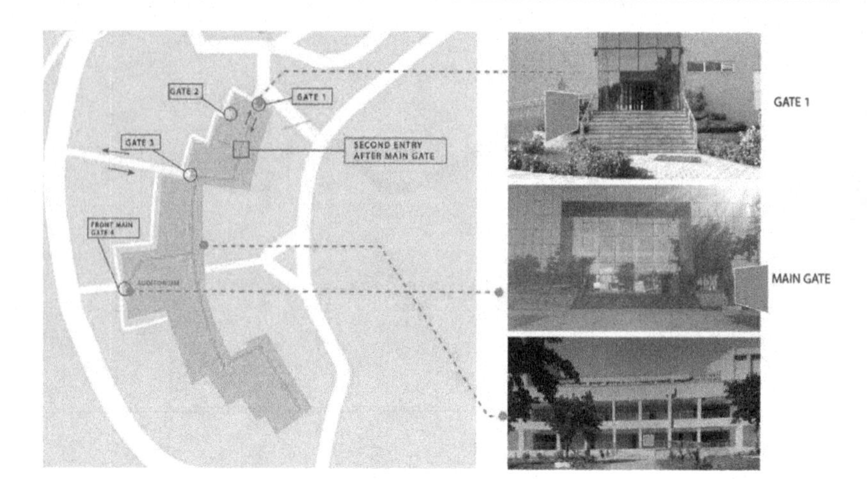

FIGURE 5.13 The map of the building where the venue is along with actual photographs of the same. As there are two entrances to the building, the same is depicted here.

FIGURE 5.14 This indicates along with the map the actual elevation view of the building and how it looks like as one move closer the venue.

photographs in the inset. The boards in the photographs indicate the position where information would be displayed for the users for giving them directions towards the venue as they move in the space.

Figures 5.15 and 5.16 indicate the path of navigation as users move closer to the venue. Reconfirmation is needed as we have seen before and this

FIGURE 5.15 This is the path of navigation.

FIGURE 5.16 The path of navigation further closer to the venue.

reconfirmation information is provided by the blue boards positioned in the path of the user's navigation and in their visual cone.

Figure 5.17 shows the map of the campus and with coding of different zones. This makes it easy for anyone coming on campus to take a decision about where to go and how to go.

Thus, while designing maps or signage, some simple visual ergonomic principles need to be followed. Maps should give users an overview of the

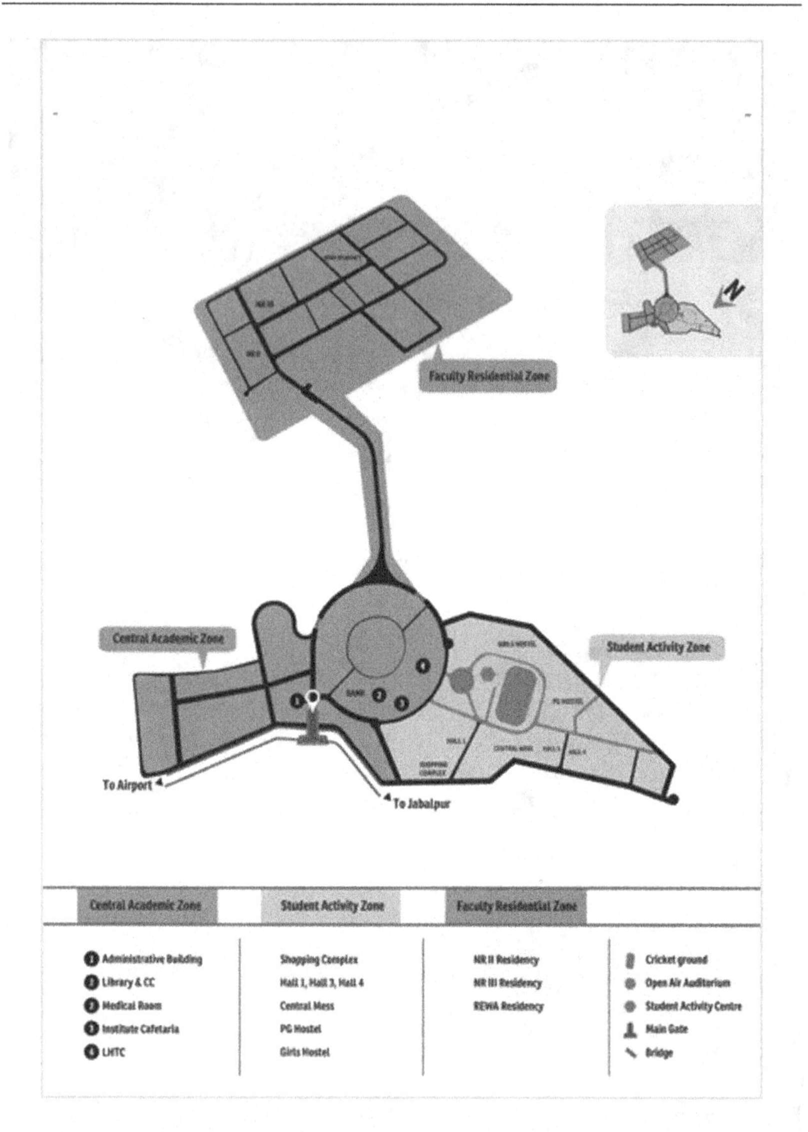

FIGURE 5.17 The map of the campus with coding for different areas.

place and the different elements or buildings in place. If supplemented by actual photographs, maps are easy to understand for common users. As users move in space with maps, there needs to be information in the space as well which should connect with the information in the map. This is what was done in this example. Both the map and the information in the space work together to help users navigate in space. Colour coding on maps helps in differentiating different areas on the campus based on different activities like sports, recreation, etc. This makes it easier for users to scan and detect without wasting time. As mentioned before that, it's always wise to provide information in a manner that is neither too short nor too long. This "trade-off" in information is necessary and is augmented through the user study.

5.6 KEY POINTS

A. Guide your users in steps.
B. Avoid long instructions.
C. Users make mistakes; you need to design in a manner that the chances are reduced.
D. Remove unnecessary information especially when it comes to filling up some information.
E. Segregate and group information meant for users and the staff.
F. Information in space hand-holds users and helps them to navigate with ease.
G. As users navigate in space, they need to know their exact location with respect to their destination and from where they started.
H. Information in space hand-holds users and should help them move inside the space and also help them come out of the space.
I. Maps and information in space should be designed in tandem and one should supplement the other.
J. Colour coding in maps is just one of the many ways in differentiating between the different zones in physical space.

5.7 PRACTICE SESSION

A. Use visual ergonomic principles in designing the survey form for collecting data from students regarding their medical symptoms after recovering from COVID-19.
B. Use visual ergonomic principles for designing a simple question paper on mathematics for standard 4 students dealing with addition and subtraction only. Students should be able to answer on the question paper itself.

C. An event is being organised at your institute. You are to design the map for the users so that they are able to come to the venue with ease.
D. Design the information system inside the academic block of your institute so that students do not have to ask anyone for different tasks like fee payment, transcripts, etc.

Exercises in Visual Ergonomics Related to Communication Design: Some Directions

6

CHAPTER 1: THE WORLD OF WONDERS, ORIGIN, AND HOW TO MOVE FORWARD

Assignment 1: Your college canteen sells different types of fruit juice in sealed bottles at one corner. The area is dull and students have a hard time identifying the type of fruit juice that is being sold. Your expertise is sought to make the space more user friendly by enhancing the communication between the products, the area, and the students who come and drink there.

Directions 1: The space needs to be analysed for proper illumination in the absence of which nothing can be seen. You need to colour the walls in a manner that it's easily identifiable from a distance and attracts the students. For attracting students, you can paint the walls with some images which could be cartoon characters of some popular faces on campus (with their due permission) with which students

DOI: 10.1201/9781003369516-6

could connect. The labelling on the fruit juice bottles should have the icon of the particular fruit so that students can identify their favourite fruit, and the price and expiry dates should be clear. These directions would ensure clear communication between the space, products, and the target users, i.e., the students.

Assignment 2: You travel from home/hostel to your design studio every day. On the way, you come across so many different pieces of information around you. These pieces of information are an attempt to make you aware of certain things, like warning against wild animals, not to venture into the forest zone, way to the cafeteria, etc. You have to list down the different visual ergonomic issues around you and the drawbacks in them that you feel need to be worked upon.

Directions 2: List down all the information along the path as you move. See what difficulties you face in understanding them at different times of the day, for example under bright sunlight, while it's raining, at night. Based on your observation, list down the visual ergonomic issues that you feel need to be incorporated. It could be proper positioning, enhancement of illumination, use of a particular colour, for example in a green environment a green signage would merge, and a different coloured board should be used so that it "pops" up from the surroundings.

CHAPTER 2: THE CODING OF VISUAL INFORMATION

Assignment 1: The symbol of the toilet for males and females in your town is a cause of concern for many as users tend to get confused about which is for which gender. As a visual ergonomist, you are to identify some visual ergonomic features and suggest changes which would reduce this uncertainty among users of both the genders.

Directions 1: Your first task should be to check for the mental model of the users which best defines the two genders, male and female. If you have to design a symbol for the third gender you need to find out the features for the same as well. After this, you have to draw the symbol with the features and ensure that you do not add too much detail as that might confuse your users. Next, take a call on the colour with specific emphasis on the background and the foreground. Lastly, decide where you can put it so that it falls within the visual cone of all the users and within the users' "expected quadrant".

Assignment 2: You are to give your expert advice from a visual ergonomics perspective as to how exactly should a street name be written including the font, size, and the background and foreground colour. The problem is that the place is very dimly lit at night.

Directions 2: You need to select a font family with which your users are familiar with. The next call that you have to take is whether you should use all capital or upper and lower case. In this case, use of upper and lower case would facilitate easy reading as it's a street's name. The biggest concern is that the place is dimly lit at night. In that case, a dark background with light text would be a better option but given the fact that the name of the street is not too long.

CHAPTER 3: TYPOGRAPHY AND COLOUR

Assignment 1: You have to design the layout of a newspaper for school-going children of standards 8 to 10. The layout and design of the newspaper should be such that children should find it interesting and thus develop a reading habit once again. Your expertise as a visual ergonomist is solicited. You will have with you other editorial staffs who will be helping you with selecting the news appropriate for these young people.

Directions 1: The target users are children which gives us an impression of using coloured typography. Be very careful while you use coloured typography because overdoing it would spoil the paper and make it very confusing to the users. Select a familiar font family. Some of the headers could be coloured but try to restrict yourself to two to three colours and not more. If you are adding comic strips to the newspaper "make sure they run as per the population stereotype."

Assignment 2: You have been asked by a company to design the textual material for an exhibition hall. In this hall, there would be a number of sculptors and each needs to be described in short with the name of the artist and the meaning of the art piece.

Directions 2: First look into the illumination level available in the hall. Use typography in a manner that it's a familiar font. Next, decide upon the background and text, whether light text on a dark background or dark text on a light background, depending on the illumination.

Though justification of the text looks good, it creates unequal space between the words which reduces the speed of reading. Always be precise and avoid long text as users have other exhibits to see.

CHAPTER 4: VISUAL ERGONOMICS IN EMERGENCY SITUATIONS AND FOR THE CHALLENGED AND ELDERLY

Assignment 1: You have been asked to design the information system inside a train compartment in the event of smoke inside so that passengers are able to follow the information and navigate themselves out to safety.

Directions 1: In case of smoke in a space, you should use larger fonts and thicker letters with very high contrast between the background and the foreground. More important is to refrain from using textual information, and if at all, should be restricted. Your text needs to be loud and you can think of using some capital letters for achieving that. Lastly, after you design, you need to simulate the environment and test it out on a few users.

Assignment 2: You have to design the labelling for a nasal drop meant for the elderly. The elderly need to read it carefully and then use it.

Directions 2: If you are designing for the elderly, factor in the problems that they face in old age like cataracts, blue colour blindness, and difficulty in seeing fine print. If you plan to use visuals, make sure it's in tandem with their mental model. Make sure your textual information is as brief as possible and has high contrast.

CHAPTER 5: VISUAL ERGONOMICS OF INFORMATION COLLECTION AND DISSEMINATION TO USERS

Assignment 1: Use visual ergonomic principles in designing a form for the children of standard 8 wherein they would be mentioning their career choice with reasons. This form would be later used by career counsellors to guide them further.

Directions 1: You are dealing with children so focus on using more visuals and less textual material. Give them directions on how and what they should write. Think of using all caps and upper and lower cases for information where emphasis is needed or not needed. Make sure the form does not take much time to complete; else, they will get bored.

Assignment 2: You have to use visuals and no text for guiding users (including the illiterate) in a government hospital to the radiology department from the reception. Make sure you guide the user to the department and again guide them out to the main gate.

Directions 2: Put yourself in users' shoes and decide the location of the information system or the visual. Next, decide at what intervals should they be placed and where reconfirmation needs to be given. Users are normally puzzled at crossroads; you need to help them there.

Extra Exercise (EE)

EE1: You are to decide at what minimum height a display board should be placed if the distance from the board is known. Alternately, if the height of a display board is known, from what distance do you expect users to look at it? This is a challenge that visual designers face. Another challenge they face is when designing directional signage on the road, what should be the distance between two signage? For example you are moving inside a big hospital campus. You see an arrow which points to the ophthalmology department and says "ophthalmology" department this way. After five minutes' walk, you wonder, *am I going in the right direction?* So, what should be our approach?

Directions: It has been seen through experience that users walking in space normally look for a reconfirmation after every five minutes. Considering the fact that roughly 400 metres can be covered in five minutes, you should place reconfirmatory signage after every 400 metres so that users do not need to ask anyone and can be happy that they are moving in the right direction.

Now let us come to the calculation. Figures 6.1 and 6.2 depict how this calculation is to be done.

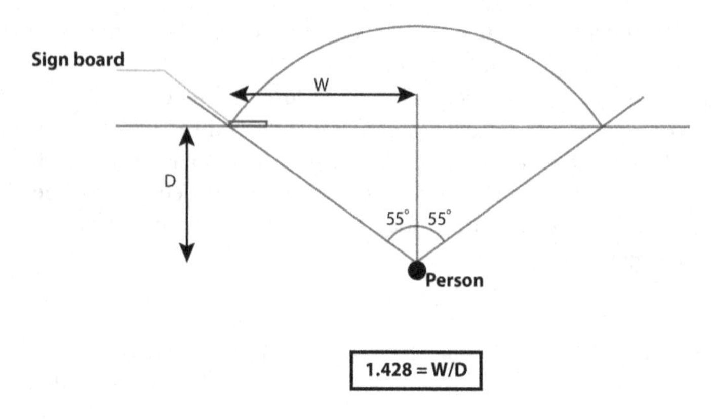

FIGURE 6.1 Calculating display placement in the horizontal cone of vision.

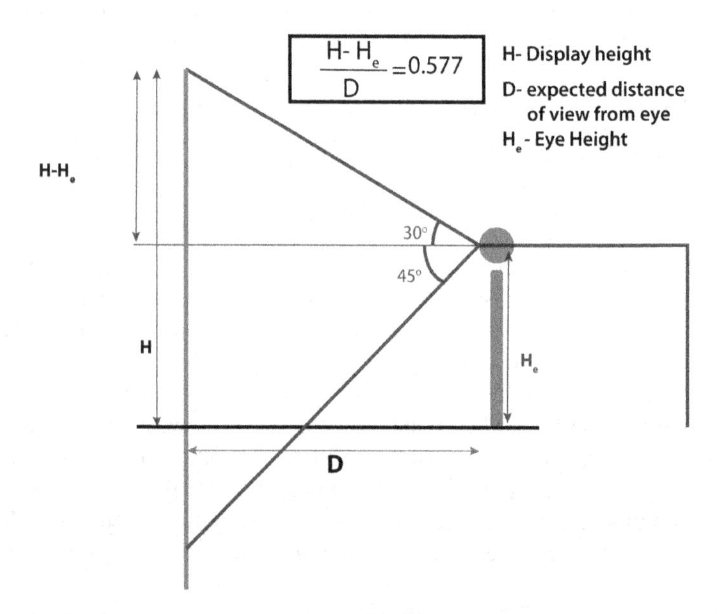

FIGURE 6.2 Calculating display placement in the vertical cone of vision.

Figure 6.1 shows the view from the top (called "plan view") where the horizontal cone of vision without moving the head is roughly 55 degrees from the centreline. So ideally, any display should lie within this visual cone for the user to see. So just by using simple trigonometry, the distance of W and D can be calculated. Tan 55 degrees is roughly 1.428. So, in the formula above 1.428 = W/D where W is the distance from the centreline (an imaginary line drawn from the space between the two eyes and parallel to the ground). So, if W is known, you can get the value of D which is the distance from the display the user should stand. On the other hand, if D is known then W can be calculated. Remember this is without moving the head. If the head moves, then the cone also moves along. So, any information should lie within this cone.

Figure 6.2 shows the user's position from the side (called "elevation"). The vertical visual field is 30 degrees above the horizontal and 45 degrees below the horizontal. This horizontal is calculated from the eye height of the users, and from the horizontal line, the 30 and 45 degree angles are drawn. So, if the user's height is known, you can calculate at what distance the display should be positioned. If the display height is given, then from what distance it can be read can also be calculated. The above formula H – He/D = 0.577 is handy. He is the eye height of the user and D is the distance of the user from the information. H is the display height. The value of tan 30 is 0.577 and tan 45 is 01. So, in the case of tan 45 (which is equal to 01), the height and distance are equal.

The calculations mentioned above are very "rough" guidelines for the placement of the display. You have to consider other factors like movement of the head in horizontal and vertical directions, movement of people, etc. Moreover, everything cannot dangle at eye level, because in that case, it becomes chaos. At what height a display can be mounted will be dictated by the range of movement of different people. This additional height of display thus mentioned would be over and above what is mentioned before. If the display is placed on a vertical surface and users stand too close, then part of the display would shift out of the visual cone. In that case, the display has to be tiled forward from the wall so as to facilitate easy scanning by the users standing close to the display. This happens in relatively small spaces where the users have to stand in close proximity to the display due to lack of space.

EE 2: You have been assigned the task of ergonomic designing of "information" at the local railway station which should make the public aware of the COVID-19 pandemic and guide them towards maintaining adequate health and hygiene during their journey. Remember a large number of passengers are unable to read or write.

Directions: As a large number of users are unable to read or write, the communication has to be through visuals with which the users are able to relate. You have to move in steps. First, decide upon the content. The content could start with an overview of COVID-19 without panicking the users. Try not to give any statistics of how many deaths have happened as that would only panic the users.

In the second step, depict COVID-19 and tell users what it is and how it's transmitted. After this, convey to your users what they should do first and then tell them what they should not do. Be clear when you suggest your users different Do's and Don'ts.

Factor in colour-blind users mainly red, green, and blue and refrain from using pure colours. The illumination level is important which would help you to decide the background and the foreground of your visuals in terms of light on dark or vice versa.

Lastly, mount your visuals so that they are in the visual cone of your users; else, all your efforts would be futile if users are not able to see or even notice them.

EE 3: A busy pharmacy in town seeks your advice as a visual ergonomic expert in solving a problem for them. They have found that they tend to make errors in dispensing medicines, especially when it comes to medicines which sound alike. The medicines come from the company so you cannot ask them to change the way the medicine packagings are designed. How would you approach this problem and give them a visual ergonomic solution?

Directions: Visual ergonomists have to work within limitations. Here, the only option is that you have to suggest ergonomic intervention on the medicine packaging that is already there. You have to differentiate the sound alike medicines. For this differentiation, you can go for colour coding (highlighting the part of the name which differentiates) or can add a sticker in which the differentiator is written in all caps. For example there are two sound alike medicines (fictitious ones!) Belpahar and Belgahar. There is every possibility that during heavy workload a user can confuse between the two. The differentiators here are "pahar" and "gahar". So, either you can highlight these two parts in two different colours (yellow and pink for example) or put a sticker and write "PAHAR" and "GAHAR" in all caps. In that case, they stand apart and alert the pharmacist. Remember we can only do some value addition to the packaging! You can think of other ways as well.

EE 4: A new residential complex is coming up in your locality. There are 15 towers, each 20 storied. Your expertise has been sought as a visual ergonomist in designing the information system in space for helping guests navigate to the desired apartment of their choice without the need to ask anyone about the direction for the same. How would you proceed?

Directions: You need to give the target users an overview at the entrance as to the location of different towers and apartments. The next stage would be guiding the users towards the desired tower with adequate confirmation and reconfirmation as they move towards the tower. Once they are close to the tower, another set of information overview is given to indicate the placement of different apartments in the respective towers. Remember that you need to guide your users once they come out of the respective apartments towards the exit gate as well. This is often not thought of.

EE 5: A roadside eatery wants you to design the menu card for them, which is to be used mainly by different types of tourists both local and global. The eatery advises you to refrain from using any textual materials as language would be a problem for many and instead focus on specific visuals which are comprehensible to all. How would you approach this problem from the visual ergonomic perspective?

Directions: When you have to use only visuals, you need to ensure that they are in tandem with the user's mental model. Here we are dealing with an international population and hence the visuals should be universal and everyone should be able to understand them. The different types of dishes need a representation of how they would look like, their ingredients, and it should show clearly vegetarian, non-vegetarian, and vegan dishes as well. You have to use colour prudently to bring out the look and feel of different dishes and this exercise has to go through a series of user testing before coming up with the final solution.

Bibliography

Lin, R., Lin, P. C., & Ko, K. J. (1999). A study of cognitive human factors in mascot design. *International Journal of Industrial Ergonomics*, *23*(1–2), 107–122.

Mayhorn, C. B., Wogalter, M. S., & Bell, J. L. (2004). Homeland security safety symbols: are we ready?. *Ergonomics in design*, *12*(4), 6–11.

Mukhopadhyay, P. (2013). Ergonomic Design Issues in Icons Used in Digital Cameras in India. International Journal of Art, Culture and Design Technologies (IJACDT), 3(2), 51–62.

Mukhopadhyay, P. (2019). *Ergonomics for the Layman: Applications in Design*. CRC Press.

Mukhopadhyay, P. (2022). *Ergonomics Principles in Design: An Illustrated Fundamental Approach*. CRC Press.

Mukhopadhyay, P., & Vinzuda, V. (2019). Ergonomic Design of a Driver Training Simulator for Rural India. In Advanced Methodologies and Technologies in Artificial Intelligence, Computer Simulation, and Human-Computer Interaction (pp. 293–311). IGI Global.

Mukhopadhyay, P., Kaur, J., Kaur, L., Arvind, A., Kajabaje, M., Mann, J., ... & Chakravarty, S. (2013). Ergonomic design analysis of some road signs in India. *Information Design Journal (IDJ)*, *20*(3), 220–227.

Ng, A. W., & Chan, A. H. (2018). Color associations among designers and non-designers for common warning and operation concepts. Applied ergonomics, 70, 18–25.

Patel, G., & Mukhopadhyay, P. (2021). Ergonomic analysis and design intervention in symbols used in hospitals in central India. *Applied Ergonomics*, *94*, 103410.

Patel, G., & Mukhopadhyay, P. (2022). Comprehensibility evaluation and redesign of safety/warning pictograms used on pesticide packaging in Central India. *Human and Ecological Risk Assessment: An International Journal*, *28*(1), 22–42.

Phillips, R. J. (1979). Why is lower case better? Some data from a search task. *Applied Ergonomics*, *10*(4), 211–214.

Phillips, R. J., & Noyes, L. (1977). Searching for names in two city street maps. *Applied Ergonomics*, *8*(2), 73–77.

Phillips, R. J., Noyes, E., & Audley, R. J. (1978). Searching for names on maps. *The Cartographic Journal*, *15*(2), 72–77.

Phillips, R. J., Noyes, L., & Audley, R. J. (1977). The legibility of type on maps. *Ergonomics*, *20*(6), 671–682.

Piamonte, D. P. T., Abeysekera, J. D., & Ohlsson, K. (2001). Understanding small graphical symbols: a cross-cultural study. *International Journal of Industrial Ergonomics*, *27*(6), 399–404.

Sanders, M. S., & McCormick, E. J. (1987). Human factors in engineering and design. McGRAW-HILL book company.

Index

Printed in the United States
by Baker & Taylor Publisher Services